AI FOR HEALTHCARE ROBOTICS

What is artificial intelligence (AI)? What is healthcare robotics? How can AI and healthcare robotics assist in contemporary medicine?

Robotics and AI can offer society unimaginable benefits, such as enabling wheelchair users to walk again, performing surgery in a highly automated and minimally invasive way, and delivering care more efficiently. *AI for Healthcare Robotics* explains what healthcare robots are and how AI empowers them in achieving the goals of contemporary medicine.

AI FOR EVERYTHING

Artificial intelligence (AI) is all around us. From driverless cars to game-winning computers to fraud protection, AI is already involved in many aspects of life, and its impact will only continue to grow in the future. Many of the world's most valuable companies are investing heavily in AI research and development, and not a day goes by without news of cutting-edge breakthroughs in AI and robotics.

The *AI for Everything* series will explore the role of AI in contemporary life, from cars and aircraft to medicine, education, fashion, and beyond. Concise and accessible, each book is written by an expert in the field and will bring the study and reality of AI to a broad readership including interested professionals, students, researchers, and lay readers.

AI for Art
Niklas Hageback & Daniel Hedblom

AI for Creativity
Niklas Hageback

AI for Death and Dying
Maggi Savin-Baden

AI for Radiology
Oge Marques

AI for Games
Ian Millington

AI for School Teachers
Rose Luckin. Karine George & Mutlu Cukurova

AI for Learning
Carmel Kent & Benedict du Boulay

AI for Social Justice
Alan Dix and Clara Crivellaro

AI for the Sustainable Development Goals
Henrik Skaug Sætra

AI for Healthcare Robotics
Eduard Fosch-Villaronga & Hadassah Drukarch

AI for Physics
Volker Knecht

For more information about this series please visit: https://www.routledge.com/AI-for-Everything/book-series/AIFE

AI FOR HEALTHCARE ROBOTICS

EDUARD FOSCH-VILLARONGA

HADASSAH DRUKARCH

CRC Press is an imprint of the
Taylor & Francis Group, an **informa** business

First Edition published 2022
by CRC Press
6000 Broken Sound Parkway NW, Suite 300, Boca Raton, FL 33487-2742

and by CRC Press
4 Park Square, Milton Park, Abingdon, Oxon, OX14 4RN

CRC Press is an imprint of Taylor & Francis Group, LLC

Library of Congress Cataloging-in-Publication Data
Names: Fosch-Villaronga, Eduard, author. | Drukarch, Hadassah, author.
Title: AI for healthcare robotics / Eduard Fosch-Villaronga, Hadassah Drukarch.
Description: First edition. | Boca Raton : CRC Press, 2022. |
Series: AI for everything | Includes bibliographical references and index.
Identifiers: LCCN 2021054783 | ISBN 9781032063300 (hardback) |
ISBN 9781032061283 (paperback) | ISBN 9781003201779 (ebook)
Subjects: LCSH: Artificial intelligence–Medical applications. |
Machine learning. | Medical informatics.
Classification: LCC R859.7.A78 F67 2022 |
DDC 610.285–dc23/eng/20211108
LC record available at https://lccn.loc.gov/2021054783

ISBN: 978-1-032-06330-0 (hbk)
ISBN: 978-1-032-06128-3 (pbk)
ISBN: 978-1-003-20177-9 (ebk)

DOI: 10.1201/9781003201779

Typeset in Joanna
by codeMantra

CONTENTS

FOREWORD

The COVID-19 pandemic has brought to light two major challenges in healthcare: (1) worldwide, access to high-quality (or any) healthcare is out of reach for a large majority of people; (2) the healthcare workforce, which was already at its breaking point before the pandemic, is bursting. Many people in healthcare robotics, the field I work in, view these challenges as opportunities to use technology to improve the state of things – increase access to care, fill care gaps, and relieve healthcare worker burnout. These are admirable goals, and while awaiting systemic change at the healthcare delivery systems and public health and policy levels, perhaps a reasonable path to pursue.

However, it is important to tread carefully and be mindful of the complex ecosystem of healthcare. Technosolutionism has a tendency to place technology before people, and conveys an inherent reductionism that suggests technological solutions will solve inordinately complex problems. I liken this framing to 'AI Snake Oil' – the hype far exceeds the state of the art, and, in some cases, can make things worse. Fosch-Villaronga and Drukarch illustrate these nuances very clearly, including real-world examples such as opaque medical diagnostic systems that yield incorrect and dangerous results and the security vulnerabilities in many surgical robotic systems. They emphasize the importance of transparency and explainability, and the criticality of extensive testing, auditing, and public scrutiny before employing robots in healthcare settings.

Another aspect of understanding complex healthcare ecosystems is how the introduction of technology will affect clinical workflow (and, therein, patient safety). My clinic-based work is primarily in acute care, including in the Emergency Department. Here, healthcare workers are overburdened, overworked, and overstressed. They are interrupted every 6 minutes and subjected to constant loud and distracting ambient noises (alarms sounding, phones ringing, overhead pagers blaring, loud talking, and, often, shouting). To introduce a robot into this setting, without appreciating its complexity and acquiring an understanding of how work is done, is a recipe for disaster. Both technology builders and healthcare leaders need to allow time to deeply understand these settings and engage in collaborative co-design processes with healthcare workers to ensure systems fit within existing work practices.

Healthcare leaders in particular should acknowledge an augmentation rather than automation framing when considering the adoption of healthcare robots to support work. As Fosch-Villaronga and Drukarch and others in the field argue, we are not now, nor will ever likely be, in a position to automate the majority of high-level, patient-facing care tasks, which require complex knowledge, intuition, problem-solving, and empathy. Instead, robots can be used to create spaces for healthcare workers to employ their irreplaceable skills as compassionate human beings, by perhaps adopting unpleasant or unsafe work. Research that helps deepen our understanding of the role of robots and their relationship with healthcare workers will help further contextualize how/if/where/when the use of robots is appropriate in healthcare.

Another major trend in healthcare is a shift from hospital-delivered care to home and community-based care. Many chronic health conditions can be managed at home, and people are eager to avoid hospitals, particularly during an ongoing pandemic. Furthermore, home-based care delivery has the potential to improve access to care. Robots have a role to play here, though their capabilities, roles, and functionalities require further research and development, particularly with regard to their autonomy, usability, and accessibility. It is also critical to consider the cost of these technologies. Worldwide, even in developed countries, people make decisions every day about buying food or medicine. If a robot-delivered intervention is successful, it is imperative to make sure people can afford to have it.

This book provides an excellent starting point for all key stakeholders – healthcare workers, patients, caregivers, technologists, researchers, policymakers, and leaders – to navigate and understand current directions in healthcare robotics, including both possibilities and pitfalls. It outlines opportunities and challenges, and directions for new avenues for research. I admire the authors' methodological investigation of the literature to help define and understand this continually evolving field, and I hope you enjoy reading the book as much as I have.

Laurel Riek, PhD
Associate Professor, Computer Science & Engineering
Associate Adj. Professor, Emergency Medicine
Director, Healthcare Robotics Lab
UC San Diego

AUTHORS

Dr. Eduard Fosch-Villaronga is an Assistant Professor at the eLaw Center for Law and Digital Technologies at Leiden University (NL), where he investigates legal and regulatory aspects of robot and AI technologies, with a special focus on healthcare. Eduard recently published the book *Robots, Healthcare, and the Law: Regulating Automation in Personal Care* with Routledge and is interested in human–robot interaction, responsible innovation, and the future of law. Eduard is the principal investigator (PI) at Propelling, a project funded by the H2020 Eurobench project that uses robot testing zones to support evidence-based robot policies (EU's H2020 Research and Innovation Program Grant Agreement No 779963). He is also the co-leader of the project Gendering Algorithms, an interdisciplinary pilot project aiming to explore the functioning, effects, and governance policies of AI-based gender classification systems. Previously, Eduard was the PI at Liaison, a project that received the H2020 COVR Award and that aimed to link robot development and policymaking to reduce the complexity in robot legal compliance (EU's H2020 Research and Innovation Program Grant Agreement No 779966). Eduard served the European Commission in the Sub-group on Artificial Intelligence (AI), connected products and other new challenges in product safety to the Consumer Safety Network (CSN) to revise the General

Product Safety directive in 2020. Previously, he was the recipient of a personal Marie Skłodowska-Curie Postdoctoral Researcher under the COFUND LEaDing Fellows at the eLaw Center for Law and Digital Technologies at Leiden University (NL) (Jan 2019-Dec 2020). He also was a postdoc at the Microsoft Cloud Computing Research Center at Queen Mary University of London (the UK, 2018) investigating the legal implications of cloud robotics; and at the University of Twente (NL, 2017) as a postdoc, exploring iterative regulatory modes for robot governance. Eduard Fosch-Villaronga holds an Erasmus Mundus Joint Doctorate (EMJD) in Law, Science, and Technology coordinated by the University of Bologna (IT, 2017), an LL.M. from the University of Toulouse (FR, 2012), an M.A. from the Autonomous University of Madrid (ES), and an LL.B. from the Autonomous University of Barcelona (CAT, 2011). Eduard is also a qualified lawyer in Spain, fluent in Catalan, Spanish, English, Italian, and French.

Hadassah Drukarch is a Research Assistant at the eLaw Center for Law and Digital Technologies at Leiden University (NL), a paralegal at the legal and public affairs consultancy firm Considerati, and the founder and host of The Law of Tech Podcast which aims to provide insights into how new technologies are impacting upon the law and legal industry. She is currently a student at the Advanced LL.M. in Law and Digital Technologies at Leiden University, The Netherlands. Moreover, she was a research trainee at eLaw in 2020, working on Healthcare Robotics, including robotic assisted surgery regulatory and liability aspects. Hadassah's work navigates between law, digital technologies, governance, business and entrepreneurship, diversity and inclusion, and human rights. She holds an LL.B. in International Business Law and an Honours College Law Program certificate from Leiden University.

INTRODUCTION

THE RISE OF AI-DRIVEN HEALTHCARE ROBOTICS

Robotics have increased productivity and resource efficiency in the industrial and retail sectors, and now there is an emerging interest in realizing a comparable transformation in healthcare (Cresswell, Cunningham-Burley, & Sheikh, 2018). Soon robots will take care of you, me, and our beloved ones in hospitals and medical centers. A robot will help a gynecologist or urologist perform crucial and life-saving surgeries; robots will help people who spent 10 years in a wheelchair walk again and climb stairs; and when you grow old, your family might bring you a social robot to keep you company and monitor you.

Robotics and artificial intelligence (AI) define a new and swiftly evolving scenario and are some of the latest promising technologies expected to increase the quality and safety of care while simultaneously restraining expenditure, especially given their success in the industrial sector (Riek, 2017; Cresswell et al., 2018). Moreover, such a transformation is encouraged by the urge to increase care quality and safety while simultaneously restraining expenditure. As such, healthcare robots are likely to be deployed to this end at an unprecedented rate (Simshaw et al., 2015). AI-driven healthcare robots allow healthcare providers to reduce their costs and are becoming particularly popular because of their increased roles and capacities (COMEST, 2017) to perform medical interventions (Nouaille et al., 2017), support impaired patients (Tucker et al., 2015), provide therapy to children (Scassellati, Admoni, & Matarić, 2012), or keep the elderly company (Broekens, Heerink, & Rosendal, 2009).

DOI: 10.1201/9781003201779-1

Within this context, AI offers yet another wave of innovation and transformation for the delivery of healthcare. AI is a new realm of science and technology which already affects many human activities at an almost endless range of societal levels, from individuals to social groups, corporations, and nations (Gómez-González, 2020). As we have experienced over the past decade, AI has expanded on a global scale at an unprecedented speed in almost every industrial, economic, and societal sector, from information technologies to commerce, manufacturing, space, remote sensing, security and defense, transport, and vehicles (Gómez-González, 2020). For instance, think about introducing AI-driven personalized social media feeds, self-driving cars, intelligent virtual assistants, or even autonomous weapon systems. More recently, AI has also presented itself in healthcare robotics, and this development has been further triggered by the outbreak of the COVID-19 pandemic, throughout which AI-related technologies have been considered to play an essential role in the fight against the virus as an ongoing international priority. This has further sparked international interest in the development and use of AI within the domain of healthcare (Aymerich-Franch & Ferrer, 2020; Khan, Siddique, & Lee, 2020).

AI is increasingly gaining a prominent presence in the domain of healthcare robotics. It augments and advances the capabilities of healthcare robots and combines the potential of disruptive advances with extraordinary benefits in medicine and healthcare. Yet while the recent advances in AI systems in medicine and healthcare present tremendous opportunities in many areas, they inevitably also raise important questions and drawbacks. This will require us to carefully consider their implementation and the possible ways in which they may affect and change our understanding of healthcare delivery and the basic concepts and definitions that have traditionally characterized this domain.

Interestingly, however, although the field of healthcare robotics and AI is very rich and extensive, it is currently still very much scattered and unclear in terms of definitions, medical and technical classifications, product characteristics, purpose, and intended use

(Fosch-Villaronga & Drukarch, 2021). As we will further explain in this book, having unclear definitions and categories in place adversely impacts, among other things, the understanding of how the legislation applies to concrete robot applications and how it hampers compliance processes and safety. This is essential to prevent unnecessary harm from occurring and to ensure public trust in the system we rely upon, especially in the challenging times we have been presented with recently. Although the use of AI and robotics in healthcare may seem distant and of a somewhat science-fiction type nature to many, it will not be long before we encounter practical examples of AI and robotics in healthcare delivery on a personal and individual level. This will inevitably change our understanding of and interaction as users and patients within the healthcare ecosystem, requiring us to grasp better what the near future will likely hold for us as humans in an increasingly tech-based environment.

WHAT IS THIS BOOK ABOUT?

This book introduces the domain of healthcare robotics and the application of AI within this context. For this purpose, we limit our focus to surgical, assistive, and service robots to rightfully match the definition of healthcare as the organized provision of medical care to individuals, including efforts to maintain, treat, or restore physical, mental, or emotional well-being. To this end, this book provides a structured overview of and further elaboration on the main healthcare robot categories now established, their intended purpose, use, and main characteristics. An elaborate insight integrating AI into the current state-of-the-art healthcare robots and the impact on the healthcare robot ecosystem complements this overview.

STRUCTURE OF THIS BOOK

This book is divided into six main chapters, preceded by an introduction and followed by a conclusion. The first chapter introduces the importance of having definitions intersecting healthcare, robotics, and AI. It thereby provides working definitions that provide a basis for understanding this complex interplay and explains how these systems are embodied, their autonomy levels, and how users interact with these systems. Chapter 2 elaborates on the introduction of AI in healthcare – explaining how this has significantly reshaped medical practice and raised questions and concerns regarding transparency and explainability – and concludes by mapping the healthcare robot ecosystem and highlighting the increasingly complex interplay between all involved actors. The remaining chapters cover the ecosystem and state of the art of healthcare robotics and the application of AI within this context, following the established categories of surgical robotics, socially assistive robots, physically assistive robots, and healthcare service robots. The book concludes with some remarks on the impact and implications of developing and deploying AI-driven robotics in the healthcare domain.

REFERENCES

Aymerich-Franch, L., & Ferrer, I. (2020). The implementation of social robots during the COVID-19 pandemic. arXiv preprint arXiv:2007.03941. https://arxiv.org/abs/2007.03941.

Broekens, J., Heerink, M., & Rosendal, H. (2009). Assistive social robots in elderly care: a review. *Gerontechnology*, 8(2), 94–103. https://doi.org/10.4017/gt.2009.08.02.002.00.

Cresswell, K., Cunningham-Burley, S., & Sheikh, A. (2018). Health care robotics – qualitative exploration of key challenges and future directions. *Journal of Medical Internet Research*, 20(7), 1–11. [e10410]. https://doi.org/10.2196/10410.

Fosch-Villaronga, E. & Drukarch, H. (2021). On healthcare robots. Concepts, definitions, and considerations for policymaking. Available at ArXiv: https://arxiv.org/abs/2106.03468.

Gómez-González, E. (2020). Artificial intelligence in medicine and healthcare: applications, availability and societal impact. Gomez Gutierrez, E. (ed.), EUR 30197 EN, Publications Office of the European Union, Luxembourg, ISBN 978-92-76-18454-6 (online), JRC120214. https://doi.org/10.2760/047666.

Khan, Z. H., Siddique, A., & Lee, C. W. (2020). Robotics utilization for healthcare digitization in global COVID-19 management. International Journal of Environmental Research and Public Health, 17(11), 3819. https://doi.org/10.3390/ijerph17113819.

Nouaille, L., Laribi, M. A., Nelson, C. A., Zeghloul, S., & Poisson, G. (2017). Review of kinematics for minimally invasive surgery and tele-echography robots. ASME Journal of Medical Devices 11(4), 040802. https://doi.org/10.1115/1.4037053.

Riek, L. D. (2017). Healthcare robotics. *Communications of the ACM*, 60(11), 68–78. https://doi.org/10.1145/3127874.

Scassellati, B., Admoni, H., & Matarić, M. (2012). Robots for use in autism research. Annual review of Biomedical Engineering, 14, 275–294. https://doi.org/10.1146/annurev-bioeng-071811-150036.

Simshaw, D., Terry, N., Hauser, K., & Cummings, M. (2016). Regulating healthcare robots: maximizing opportunities while minimizing risks. Richmond Journal of Law and Technology, 22(3).

Tucker, M. R., Olivier, J., Pagel, A., Bleuler, H., Bouri, M., Lambercy, O., … Gassert, R. (2015). Control strategies for active lower extremity prosthetics and orthotics: a review. *Journal of Neuroengineering and Rehabilitation*, 12(1), 1. https://doi.org/10.1186/1743-0003-12-1.

World Commission on the Ethics of Scientific Knowledge and Technology (COMEST) (2017). Report of COMEST on Robotics Ethics. SHS/YES/COMEST-10/17/2 REV.

1

DEFINING THE DOMAINS OF ROBOTICS AND ARTIFICIAL INTELLIGENCE

1.1 THE IMPORTANCE OF DEFINITIONS

Precise terminology has always been important. Even if definitions are not an outcome in itself but merely a single step in the long process of understanding, terms, words, and vocabulary, in general, are still the primary reference we acknowledge and to which we turn to define and understand concepts, ideas, and notions. In one of his six works, *Topics*, the Greek philosopher Aristotle already identified the importance of definitions and adequate terminology, which became a central part of his philosophy. Definitions were also an important matter for his teacher Plato and the Early Academy. In fact, concerns related to the adequacy of definitions and correct terminology are at the center of the majority of Plato's dialogues, some of which put forward methods for finding definitions for the understanding of this world. For Aristotle, a definition should be defined as 'an account which signifies what it is to be for something.' This phrase and its many variants point out a crucial element in understanding the role and importance of definitions in our understanding of the world; giving a definition is saying, of some existent thing, what it is, and not simply specifying the mere meaning of a word. Put simply: Definitions formulate their essence.

DOI: 10.1201/9781003201779-2

In general terms, definitions are called into existence to create more clarity and avoid misunderstandings when discussing a particular subject. However, not all concepts are easy to describe. For instance, consider the term *emotion*. Everyone knows what emotions are until asked to provide a definition. Likewise, there are also many definitions for what we understand as *intelligence*. Even experts do not seem to agree on what intelligence is when they are asked to define it. Something similar happens with the word *robot* or with the phrase *artificial intelligence*. Still, knowing the precise terminology is crucial, especially when we try to apply and understand the same concepts in different contexts.

This is particularly noticeable in those fields intersecting law and new technologies, where the use and meaning of words differ entirely in different contexts and according to the communities by which they are used. For instance, consider both the legal and computer science domains. In the legal field, the term *transparent* is generally defined as 'easy to perceive or detect.' However, it seems that the Oxford Dictionary also highlights that within the context of computing, this term means 'of a process or interface functioning without the user being aware of its presence.' While both fields thus extensively make use of the term *transparency*, they both understand and apply this term in completely different ways (Felzmann et al., 2019). This causes much confusion because the term as used by the former community seems to be juxtaposed to the understanding the latter has attributed to the very same concept and *vice versa*. In this particular context, this confusion could lead to developers not fulfilling the legal requirements that the law imposes with respect to transparency and explainability in domains such as data protection or artificial intelligence (AI) regulation.

Especially in today's rapidly evolving society, it is not always possible to anticipate all possible developments and to adequately define them once they present themselves. This has especially proved to be a huge cause for concern where various domains intersect with one another, such as already indicated above for the domains of technology, law, and healthcare. For these fields to interact harmoniously, it will be necessary to gain a clear understanding of and clarify the components making up the transition towards digital healthcare. This, first, calls for a clarification of some of the general terminologies at the center of this complex interaction: robots and AI, autonomy levels, and human–robot interaction (HRI). Based on this understanding, Chapter 2 continues to delve deeper into the domain of healthcare robotics.

1.2 GUIDING YOUR WAY IN THE WORLD OF HEALTHCARE ROBOTICS: GENERAL TERMINOLOGY

1.2.1 ROBOTS

Similar to what we identified with *emotions*, it seems that everyone knows what a robot is, until asked to give a definition (Fehr & Russell, 1984; SPARC, 2015; Simon, 2017). Etymologically speaking, the word *robot* derives from the archaic Czech word *robota*, and means 'forced, serf labor.' The word robot was introduced into the English vocabulary for the first time after the play 'Rossumovi Univerzální Roboti' (Rossum's Universal Robots, R.U.R.), written by Karek Čapek in 1920, and staged in New York in 1922 (Čapek, 2004). R.U.R. was a play, for which Čapek invented the word robot, and involved a scientist named Rossum who discovered the secret of creating human-like machines which he then produces and distributes worldwide through a newly established factory. At the same time, another scientist decides to make the robots more human, which he does by gradually adding such traits as the capacity to feel pain. Roboti were human-like machines that were supposed to serve humans and do their tedious work but eventually came to dominate them completely. With his play, Čapek wanted to criticize the mechanization of human workers as a result of the industrial revolution (Horáková & Kelemen, 2003). Today, the Oxford dictionary reads 'a machine resembling a human being and able to replicate certain human movements and functions automatically.'

This definition has taken different shapes and forms in different communities, especially in the engineering one. For more technical definitions, some authors use classical definitions such as 'machines, situated in the world, that sense, think and act.' Others, like the International Standard Organization (ISO), define robots as 'actuated mechanism(s) programmable in two or more axes with a degree of autonomy, moving within its environment, to perform intended tasks.' (ISO 8373:2012).

Although naming a thing is to acknowledge its existence as separate from everything else that has a name (Popova, 2015), there are nonetheless many terms that do not have a legal definition. Some legal scholars in the United States (U.S.) have defined a robot as a 'constructed system that displays both physical and mental agency but is not alive in the biological sense' (Richards & Smart, 2016).The Japanese Electric Machinery Law (1971) defined an industrial robot as an 'all-purpose machine, equipped with a memory device and a terminal device (end-effector), capable of rotation and of replacing human labour by the automatic performance of movements' (Mathia, 2010).

In this respect, Bertolini and Palmerini gave a relevant definition in the context of the EU Robolaw project in 2014:'A machine, which (1) may be either provided of a physical body, allowing it to interact with the external world, or rather have an intangible nature – such as a software or program – (2) which in its functioning is alternatively directly controlled or simply supervised by a human being, or may even act autonomously in order to (3) perform tasks, which present different degrees of complexity (repetitive or not) and may entail the adoption of not predetermined choices among possible alternatives, yet aimed at attaining a result or provide information for further judgment, as so determined by its user, creator or programmer, (4) including but not limited to the modification of the external environment, and which in so doing may (5) interact and cooperate with humans in various forms and degrees' (Bertolini & Palmerini, 2014).

In 2017, the European Parliament (EP) called on the European Commission (EC) to take regulatory action – in the form of a Directive and on the basis of arts. 225 and 114 TFEU – with respect to robots and AI (European Parliament, 2017). In it, the EP acknowledged that, at that time, there was no EU definition for 'cyber-physical systems,' 'autonomous systems,' and 'smart autonomous robots.' Accordingly, the EP recommended the EC establishing a definition for such systems taking into consideration the following characteristics:

1. The acquisition of autonomy through sensors and/or by exchanging data with its environment (inter-connectivity) and the trading and analyzing of those data.
2. Self-learning from experience and by interaction (optional criterion).
3. At least a minor physical support.
4. The adaptation of its behavior and actions to the environment.
5. And the absence of life in the biological sense.

The EC defined robots as 'AI in action in the physical world.' They also called it embodied AI. For them

> a robot is a physical machine that has to cope with the dynamics, the uncertainties and the complexity of the physical world. Perception, reasoning, action, learning, as well as interaction capabilities with other systems are usually integrated in the control architecture of the robotic system.

To make it simpler, for this book we define a *robot* as 'a movable machine that performs tasks either automatically or with a degree of autonomy' (ISO 8373:2012; Richards & Smart, 2016; Fosch-Villaronga & Millard, 2019). Examples of robots include robotic manipulators to pick boxes or help build cars, self-driving cars, trucks or vans, drones, socially assistive robots, robotic vacuum cleaners, or conversational agents.

1.2.2 ARTIFICIAL INTELLIGENCE

Humans have long imagined other types of lives and intelligence. A famous example of this can be traced back to the 1818 novel *Frankenstein* written by English author Mary Shelley. This well-known novel tells the story of Victor Frankenstein, a young scientist who creates a sapient creature in an unorthodox scientific experiment, thereby imagining the creation of new types of life and inspiring many generations to come in following similar paths. AI is a concept from computer science, based on statistics, and is tightly related to pattern recognition. Although it is difficult to pinpoint, the roots of AI can probably be traced back to the second half of the 20th century, and although the emergence and further development of AI have brought society new hope and massive benefits, increasingly we are being faced with their dangers too. To quote Mary Shelly's Frankenstein in this regard,

> I had worked hard for nearly two years, for the sole purpose of infusing life into an inanimate body. For this I had deprived myself of rest and health. I had desired it with an ardor that far exceeded moderation; but now that I had finished, the beauty of the dream vanished, and breathless horror and disgust filled my heart.
>
> Shelley (2018)

It was in 1942 that the famous American Science Fiction writer Isaac Asimov responded to these concerns through his short story, *Runaround*, which revolves around the Three Laws of Robotics:

1. A robot may not injure a human being or, through inaction, allow a human being to come to harm.
2. A robot must obey the orders given to it by human beings except where such orders would conflict with the First Law.
3. A robot must protect its own existence as long as such protection does not conflict with the First or Second Laws.

This work has inspired many generations of scientists in robotics, AI, and computer science and is still inspiring work to this day. At roughly the same time, but over 3,000 miles away, the English mathematician Alan Turing told the gathering at the 1947 meeting of the London Mathematical Society that he had conceived a computing machine that could exhibit intelligence (Alan Turing, 1947). Turing thought nevertheless that whether machines could think or not was 'too meaningless to deserve discussion,' at the time. However, he believed that at the end of the 20th-century people could talk about machines thinking without being contradicted (Turing, 1950). Nowadays, and not far away from this vision, robots are considered machines, situated in the world that sense, think, and act (Bekey, 2012). The phrase *Artificial Intelligence* was then officially coined in 1956, after which the field experienced many ups and downs (Haenlein & Kaplan, 2019). Due to the rise of Big Data and improvements in computing power, many advancements are often referred to as 'AI.' These include machine learning, image recognition, smart speakers, and self-driving cars – all of which is possible due to advances in AI – and without which life as we know it would become unrecognizable.

The EC defined AI in the Communication COM/2018/237 named 'Artificial Intelligence for Europe.' They referred to AI as 'systems that display intelligent behavior by analyzing their environment and taking actions – with some degree of autonomy – to achieve specific goals' (European Commission, 2018). The term AI contains an explicit reference to the notion of intelligence. However, as seen earlier, intelligence remains a vague concept, even though it has been investigated minutely by many disciplines, including psychology, biology, and neuroscience. AI researchers mostly use the concept of rationality, which refers to the ability to choose the most optimal action to achieve a specific goal, given particular criteria and the available resources. Although rationality is not the only ingredient in intelligence, it is a significant part of it and forms the basis of machine learning, of AI based on the idea that systems can learn from data, identify patterns, and make decisions with minimal human intervention. According to them,

> AI-based systems can be purely software-based, acting in the virtual world (e.g. voice assistants, image analysis software, search engines, speech and face recognition systems) or AI can be embedded in hardware devices (e.g. advanced robots, autonomous cars, drones or Internet of Things applications)
>
> European Commission (2018)

1.2.3 EMBODIMENT, AUTONOMY LEVELS, AND HUMAN-ROBOT INTERACTION

The physical embodiment of a robot confines its capabilities and distinguishes it from mere virtual agents (Fosch-Villaronga, 2019). The embodiment of the robot plays a central role in many contexts. For instance, a robot may need to have a highly sophisticated embodiment to operate a person. If used for therapies, the robotic platform needs to be appealing and entertaining for children so they can really take advantage of what robot therapy has to offer (Tapus, Tapus, & Mataric, 2009).

Medical robots' embodiment and capabilities differ vastly across surgical, physically/socially assistive, or serviceable contexts (Fosch-Villaronga et al., 2021). The involved HRI is also very distinctive. For example, socially assistive robots interact with users socially, performing a task for the user using words and communication capabilities. However, there is close to zero contact with the user at the physical level. Physically assistive robots (like lower-limb exoskeletons), on the contrary, work toward a seamless integration with the physical user's movement. They are often attached to the user's body and help them walk or move around. Surgical robots are collaborative robots that extend the surgeon's abilities. For that, the robot embodiment needs to be very precise, include cameras to replace the doctor's eyes, and also incorporate mechanical pieces to help perform the surgery.

Autonomy comes from the Greek *autos* ('self') and *nomos* ('law'), constitutes a significant and essential aspect of contemporary robotics and HRI, and refers to 'the quality or state of being self-governing'

(Merriam-Webster). As such, the term 'robot autonomy' refers to a robot's capability to execute specific tasks based on current state and sensing without human intervention (ISO 8373:2012), and within this a number of levels of autonomy can be distinguished which define the robot's progressive ability to perform particular functions independently. Robotic autonomy varies extensively across different robot types, and it ranges from teleoperation to fully autonomous systems, influencing how humans and robots may interact between them. These ascending levels constitute a significant and essential aspect of contemporary robotics and HRI, and understanding this complex interaction becomes particularly important with sensitive domains, such as the domain of healthcare. The autonomy varies extensively across different robot types. It ranges from teleoperation to fully autonomous systems, influencing how humans and robots may interact between them. For the automotive industry, the Society of Automotive Engineers established different automation levels to clarify the progressive development of automotive technology. However, no universal standards have been defined for medical robots yet (Fosch-Villaronga et al., 2021).

1.2.4 CLOUD ROBOTICS

Ibana was probably the first to anticipate cloud robotics when in 1997 he wrote:

> A remote-brained robot does not bring its own brain with the body. It leaves the brain in the mother's environment, by which we mean the environment in which the brain's software is developed, and talks with it by wireless links
>
> Ibana (1997)

A couple of decades later, cloud computing is now mainstream, and the boundaries between 'cyber' and 'physical' are becoming increasingly blurred. Cloud computing essentially involves the use of computing resources over a network, typically the Internet, scalable according to demand. More particularly, the National Institute of Standards and Technology (NIST) has defined cloud computing as a 'model for enabling ubiquitous, convenient, on-demand network access to a shared pool of configurable computing resources [...] that can be rapidly provisioned and released with minimal management effort or service provider interaction' (Mell & Grance, 2011).

In 2010, Kuffner (2010) described these advantages of using cloud computing services in robotics as a means for providing a shared knowledge database, offloading heavy computing tasks, and creating a reusable library of skills or behaviors that map to perceived complex situations. In the same year, others were also announcing cloud-computing frameworks for service robotics (Arumugam et al., 2010). The concept of *cloud robotics* has since been extended to cover 'any robot or automation system that relies on data or code from a network to support its operation, i.e. where not all sensing, computation and memory is integrated into a single standalone system' (Kehoe et al., 2015).

Roboticists with requirements to process large quantities of data now have ready access to cloud robotic platforms which can greatly facilitate access to relevant resources, information, and communications (Hu et al., 2012). The RoboEarth project (2010–2014) developed a 'Cloud Robotics infrastructure, which includes everything needed to close the loop from robot to the cloud and back to the robot.' The catalyst for the project was the assumption that (at that time) *near future* robots would need to 'reliably perform tasks beyond their explicitly pre-programmed behaviours and quickly adapt to the unstructured and variable nature of tasks'; something unlikely without a cloud platform.

RoboEarth (2011) demonstrated that the use of a cloud system could create an environment where robotics knowledge and information can be shared to enhance robot performance and to enable knowledge sharing independently of robotics architecture. In addition, Waibel et al. (2011) argued that cloud may also facilitate component reuse across different systems and developers, and the leveraging of expertise about the usage, robustness, and efficiency of components (Qureshi & Koubâa, 2014). This approach has since been used to make standalone robots outperform their previous capabilities, for example by engaging better with children, or assisting the elderly in a much more natural way (Navarro et al., 2013; Park & Han, 2016; Rodić et al., 2016).

REFERENCES

Arumugam, R. et al. (2010). DAvinCi: A Cloud Computing Framework for Service Robots. In *IEEE International Conference on Robotics and Automation (ICRA)*, IEEE, Anchorage, AK, USA, pp. 3084–3089. doi: 10.1109/ROBOT.2010.5509469. https://ieeexplore.ieee.org/abstract/document/5509469?casa_token=bwGD012mvF8AAAAA:n8w-QAm0OKquULw6Yplig80UoBCuIDcDwgV4dkewDrxeFDkioZ0-IusaMRhpiL3hcSfKJ8qFtDw.

Bekey, G. A. (2012). Current trends in robotics: Technology and ethics. In Lin, P., Abney, K. and Bekey, G. A. (Eds.). (2012). *Robot ethics: The ethical and social implications of robotics*, MIT Press, Cambridge, MA. pp. 17–34.

Bertolini, A., & Palmerini, E. (2014). Regulating Robotics: A Challenge for Europe. *Legal Affairs Committee, Upcoming Issues of EU Law*, European Parliament, Brussels., Directorate-General for Internal Policies, Bruxelles. pp. 1–209. Retrieved from http://www.europarl.europa.eu/document/activities/cont/201409/20140924ATT89662/20140924ATT89662EN.pdf, 2014 (Accessed 31 March 2022).

Čapek, K. (2004). *RUR (Rossum's Universal Robots)*. Penguin, pp. 1–62.

Communication from the Commission to the European Parliament, the European Council, the Council, the European Economic and Social Committee and the Committee of the Regions. Artificial Intelligence for Europe COM/2018/237_final. Retrieved from https://ec.europa.eu/transparency/regdoc/rep/1/2018/EN/COM-2018-237-F1-EN-MAIN-PART-1.PDF.

Directive 2009/72/EC of the European Parliament and of the Council of 13 July 2009 concerning common rules for the internal market in electricity.

European Parliament resolution of 16 February 2017 on civil law rules on robotics (2015/2103(INL).

Fehr, B., & Russell, J. A. (1984). Concept of emotion viewed from a prototype perspective. *Journal of Experimental Psychology: General*, 113(3), 464.

Felzmann, H., Fosch-Villaronga, E., Lutz, C., & Tamò-Larrieux, A. (2019). Transparency you can trust: Transparency requirements for artificial intelligence between legal norms and contextual concerns. *Big Data & Society*, 6(1), 2053951719860542.

Fosch-Villaronga, E. (2019). *Robots, Healthcare, and the Law: Regulating Automation in Personal Care*. Routledge.

Fosch-Villaronga, E., & Millard, C. (2019). Cloud robotics law and regulation: Challenges in the governance of complex and dynamic cyber-physical ecosystems. *Robotics and Autonomous Systems*, 119, 77–91. https://doi.org/10.1016/j.robot.2019.06.003.

Fosch-Villaronga, E., Khanna, P., Drukarch, H., & Custers, B. H. (2021). A human in the loop in surgery automation. *Nature Machine Intelligence*, 3(5), 368–369. https://doi.org/10.1038/s42256-021-00349-4.

Haenlein M, Kaplan A. (2019). A brief history of artificial intelligence: On the past, present, and future of artificial intelligence. *California Management Review*, 61(4), 5–14. https://doi.org/10.1177/0008125619864925.

Horáková, J., & Kelemen, J. (2003). Čapek, Turing, von Neumann, and the 20th century Evolution of the Concept of Machine. In *Proceedings of International Conference in Memoriam John von Neumann*. pp. 121–135.

Hu, G. et al. (2012). Cloud robotics: Architecture, challenges and applications. *IEEE Network*, 26(3), 21–28.

Ibana, M. (1997). Remote-Brained Robots. *Proceedings of the 15th International Joint Conference on Artificial Intelligence* (IJCAI 97). pp. 1593–1606.

ISO 8373:2012 Robots and Robotic Devices – Vocabulary defines terms used in relation with robots and robotic devices operating in both industrial and non-industrial environments.

Kehoe, B. et al. (2015). A survey of research on cloud robotics and automation. *IEEE Transactions on Automation Science and Engineering*, 12(2), 398–409.

Kuffner, J. (2010). Cloud-Enabled Humanoid Robots. *Humanoids2010 Workshop "What's Next"*. Google Research. The Robotics Institute, Carnegie Mellon University.

Mathia, K. (2010). *Robotics for Electronics Manufacturing: Principles and Applications in Cleanroom Automation*. Cambridge University Press, p. 8.

Mell, P. and Grance, T. (2011). *The Nist Definition of Cloud Computing*. Special Publication 800-145, September 2011, p. 2.

Merriam-Webster, "*Definition of Autonomy*" (2020). Merriam-Webster.com. Retrieved from https://www.merriam-webster.com/dictionary/autonomy.

Navarro, J., et al. (2013). A Cloud Robotics Architecture to Foster Individual Child Partnership in Medical Facilities. In *Cloud Robotics Workshop in 26th IEEE/RSJ International Conference on Intelligent Robots and Systems*, IEEE, Tokyo, 2013, pp. 1–6. http://www.roboearth.org/wp-content/uploads/2013/03/final-11.pdf.

Park, I. W., and Han, J. (2016). Teachers' views on the use of robots and cloud services in education for sustainable development. *Cluster Computing*, 19(2), 987–999.

Popova, M. (2015). How Naming Confers Dignity Upon Life and Gives Meaning to Existence. *Brain Pickings*.

Preamble to the Constitution of WHO as adopted by the International Health Conference, New York, 19 June–22 July 1946; signed on 22 July 1946 by the representatives of 61 States (Official Records of WHO, no. 2, p. 100) and entered into force on 7 April 1948. The definition has not been amended since 1948.

Qureshi, B., & Koubâa, A. (2014). Five traits of performance enhancement using cloud robotics: A survey. *Procedia Computer Science*, 37, 220–227.

Richards, N. M., & Smart, W. D. (2016). How should the law think about robots? In Calo, R., Froomkin, A. M., & Kerr, I. (Eds.). *Robot law*. Edward Elgar Publishing. pp. 3–24.

RoboEarth (2011). "What is RoboEarth?" Retrieved from roboearth.ethz.ch/.

Robotics (2020). Multi-Annual Roadmap for Robotics in Europe. *Call 2 ICT 24 Horizon 2020*, SPARC, 2015. p. 287.

Rodić, A. et al. (2016). Development of Human-Centered Social Robot with Embedded Personality for Elderly Care. In *New Trends in Medical and Service Robots*. Springer International Publishing. pp. 233–247.

Shelley, M. (2018). *Frankenstein: The 1818 Text*. Penguin.

Simon, M. (2017). "What is a robot," In Simon, M. (Ed.) *Wired*. Last modified August 27, 2017. https://www.wired.com/story/what-is-a-robot/.

Tapus, A., Tapus, C., & Mataric, M. (2009). The Role of Physical Embodiment of a Therapist Robot for Individuals with Cognitive Impairments. In *RO-MAN 2009-The 18th IEEE International Symposium on Robot and Human Interactive Communication*. IEEE. 103–107.

The Ottawa Charter for Health Promotion (1986). Retrieved from https://www.euro.who.int/__data/assets/pdf_file/0004/129532/Ottawa_Charter.pdf.

Turing, A. (1947). "*Lecture to the London Mathematical Society*", 20 February 1947. Retrieved from https://www.vordenker.de/downloads/turing-vorlesung.pdf.

Turing, A. (1950). Computing machinery and intelligence. *Mind*, *LIX*(236), 433–460.

Waibel, M. et al. (2011). A World Wide Web for Robots. *IEEE Robotics & Automation Magazine*.

2

DEFINING HEALTHCARE ROBOTICS

2.1 HEALTHCARE ROBOTICS

It should come as no surprise that robots have not only become routine in the world of manufacturing and other repetitive labor tasks. They have also penetrated other fields, including healthcare, where they are used within entirely different environments and for until recently unfamiliar tasks, involving direct interaction with human users, in the surgical theater, the rehabilitation center, and the family room (Mataric, Okamura, & Christensen, 2008). Research has pointed out that an estimated 20% of the world's population experiences difficulties with physical, cognitive, or sensory functioning, mental health, or behavioral health either temporary or permanent, acute or chronic, and subject to change throughout one's lifespan (Riek, 2016). A significant number of these individuals experience severe difficulties with so-called Activities of the Daily Living (ADL) (Riek, 2017), which refer to the level of independence of a person and include activities and concepts such as washing, toileting, dressing, feeding, mobility, and transferring.

Health is an outcome, a state of being, which has always been highly valued and prioritized within society. Being healthy is a 'resource of living,' allowing people to function and participate in the wide range of activities that make up our society. However,

DOI: 10.1201/9781003201779-3

despite playing such a fundamental role in our everyday functioning and participation in society, the term health is often used without a clear understanding of what it exactly entails. Similarly, many definitions of care have been established, some very broad and some very specific. While an extensive elaboration on the concepts of health and care go far beyond the scope of this book, to gain a good understanding of the role of robotics and AI in the healthcare domain it is important to make a couple of brief remarks in this regard.

The concept of healthcare acts in accordance with some earlier approaches, being 'the organized provision of medical care to individuals, including efforts to maintain, treat, or restore physical, mental, or emotional well-being.' As such, healthcare encompasses the maintenance or improvement of health via the prevention, diagnosis, treatment, recovery, or cure of disease, illness, injury, and other physical and mental impairments in people as delivered by health professionals and allied health fields. It generally covers primary, secondary, and tertiary care, as well as public health, and research has attempted to point out specific aspects which can form a basis for evaluating care – structure, process, and outcome – in both a qualitative and quantitative sense (Donabedian, 1988; Sætra, 2020).

Especially in recent decades, an increasing interest in healthcare digitalization has been identified. Robotics have increased productivity and resource efficiency in many areas, including the industrial and retail sectors and agriculture and farming. Self-driving cars, autonomous vacuums, cow-milking and pepper-planting robots, self-service checkouts at grocery stores, autonomous weapon systems, drones, and virtual embodied tour guides – all are made possible thanks to rapid advancements in robotics. Seeing such profitable gain from the employment of robots and the use of artificial intelligence (AI), now there is an emerging interest in realizing a comparable transformation in the healthcare sector (Poulsen, Fosch-Villaronga, & Burmeister, 2020). Robotics and AI are some of the latest promising technologies expected to increase the quality and safety of care while simultaneously restraining expenditure and, recently, reducing human contact too. Healthcare robots are likely to be deployed at an unprecedented rate due to their reduced cost and increasing capabilities such as carrying out medical interventions, supporting biomedical research and clinical practice, conducting therapy with children, or keeping the elderly company.

Recent developments in healthcare robotics have fundamentally changed how the medical and healthcare environment's functioning is perceived within society, and the societal drivers for improved healthcare that can be addressed by robotic technology, broadly, lie in two categories: the wish to broaden access to healthcare and to improve prevention and patient outcomes (Cresswell, Cunningham-Burley, & Sheikh, 2018). In this sense, advances in robotics have shown to have clear potential for stimulating the development of new medical treatments for a wide variety of diseases and disorders, for improving both the standard and accessibility of care, for enhancing patient health outcomes, and for filling quantitative care gaps, supporting caregivers, and aiding healthcare workers (Kim, Gu, & Heo, 2016). As such, many children under the autism spectrum disorder (ADS) can now do therapy with social robots, such as the LuxAI, in which they can practice daily life skills at the cognitive, language, social, and emotional level at home.

Even more so, the spectrum of robotic system niches in medicine and healthcare currently spans a wide range of environments, including nursing homes, hospitals, and the homes of users; user populations, including children, youth, and older adults; and interaction modalities, including physical and social human–robot interaction. Compare our current healthcare environment to that of approximately 20 years ago, and the benefits of robotics and digitalization in healthcare thus instantly become evident. Right now, it is already possible to provide physical rehabilitation for patients recovering from a stroke or with spinal cord injury, and help wheelchair patients walk back again; conduct surgeries with the help of tele-operated robots; and get blood and organs for transplant delivered in a very short time thanks to drone technology – something unimaginable before.

In line with the emphasis placed on the importance of definitions and terminology in the previous chapter, here too it is of essential importance to define the concept of healthcare robots in order to understand their role and potential in today's healthcare environment. In 2008, the European Foresight Monitoring Network (EFMN) defined *healthcare robots* as 'systems able to perform coordinated mechatronic actions (force or movement exertions) based on processing information acquired through sensor technology, to support the functioning of impaired individuals, medical interventions, care and rehabilitation of patients and also individuals in prevention programs.' Such robots encompass varying degrees of autonomy and broadly include affiliated technology, including sensor systems, algorithms for processing data, and cloud services (Riek, 2016; Fosch-Villaronga, & Millard, 2019).

It is this combination of factors that led the Policy Department for Economic, Scientific, and Quality of Life Policies of the European Parliament to identify 'the most interesting applications of healthcare robots,' which include robotic surgery, care, and socially assistive robots, rehabilitation systems, and training for healthcare workers (Dolic, Castro, & Moarcas, 2019). Moreover, analysis of past and current performance of robotics within the field of healthcare have proved that there is an incredible opportunity for robotics technology within the healthcare domain; they may help fill care gaps, aid healthcare workers, and may be used for physical and cognitive rehabilitation, surgery, telemedicine, drug delivery, and patient management (Riek, 2017). Contemporarily and over these years, fears and reluctance traditionally associated with robots and AI for healthcare as seen in previous Eurobarometers seem to have been set aside, especially as they have become strong allies in the fight against the coronavirus. In this respect, many countries started to explore the usefulness of these technologies throughout the COVID-19 pandemic regarding support and assistance in socially distant contexts.

In this book, we, therefore, focus on three main categories of healthcare robots, namely: surgical robots, assistive robots, and

Table 2.1 Healthcare Robot Categories and Definitions (Fosch-Villaronga & Drukarch, 2021)

Healthcare Robot Categories		Definition
Surgical robots		Service robots supporting surgeons during surgical procedures.
Assistive robots	Socially assistive robots	Service robots assisting users through social interaction.
	Physically assistive robots	Service robots supporting users through physical interaction.
Healthcare service robots		Service robots in a healthcare setting performing tasks useful to the facility and the medical staff.

healthcare service robots (Fosch-Villaronga & Drukarch, 2021) (see Table 2.1). Moreover, we distinguish between physically assistive robots (PAR) and socially assistive robots (SAR) within the context of assistive robots. Table 2.1 provides definitions for these different healthcare robot categories:

In this book, we will talk in more detail about each of these healthcare robots, their capabilities, and their applications to help the reader get an idea of what the state of the art of these technologies is. By doing so, we will distinguish those robots that are currently being deployed in practice, those that are currently being developed and tested and promise to enter the mass market soon, and those robots that are still very much in the realm of fiction. For instance, in 2012, there was a movie called *Robot and Frank* that explained the story of an assistant robot with caring and nursing capabilities for older adults. The robot pictured in the movie has the capacity to clean; help in daily activities such as shaving, cooking, and even gardening; and also work as a sort of motivational coach – even if such a healthy behavior clashes with the free will of the patient. Although many researchers are working toward having personal assistants that work in such a way, the current state of the art is not as advanced as shown in the movie. A more real example would be the European project 'Giraff+,' which is a tele-operated robotic platform that aims at monitoring the elderly at home continuously. Although the robot can move on wheels around the house, doctors can communicate with the patient via a camera, speakers, and microphones, and the sensors around the house provide detailed information about the daily activities of the user, the robot cannot perform nor as many tasks as portrayed in the above movie nor with the seamless and perfect integration into the user's daily life as in the movie.

In this book, we also cover the tremendous possibilities that AI has to offer for healthcare robotics. Not all healthcare robots incorporate AI, though. AI for healthcare robotics is a design choice to afford learning and other capabilities for robots but it does not always need to be present. AI for healthcare robotics thus falls under a spectrum of possibilities that depend much on the task to be performed, the outcome to be achieved, and whether AI will enable a more efficient, more quickly, and safer robot performance. This is similar to their embodiment: some robots may need a simple body (e.g., a speaker to provide oral information) or others more complex (e.g., a surgery robot to perform surgery) to complete a task (Fosch-Villaronga & Millard, 2019). In the next section, we focus on the possibilities that AI brings to healthcare.

2.2 ARTIFICIAL INTELLIGENCE FOR HEALTHCARE

Following the introduction of AI in the second half of the 20th century, researchers and developers increasingly began to recognize that AI systems could significantly benefit the healthcare domain, and especially in the 1980s and 1990s, it was argued that in healthcare such technology must be designed to accommodate the absence of perfect data and build on the expertise of physicians (Randolph & Miller, 1994). Over time, it became evident that AI, which involves machine learning and natural language processing, serves exceptionally well in revolutionizing any knowledge-intensive sector, including in particular the healthcare sector (Garbuio & Lin, 2019; Lee & Yoon, 2021).

Different medical domains previously reserved for human experts are increasingly augmented or transformed completely thanks to the integration of AI in clinical practice, including disease diagnosis, automated surgery, patient monitoring, translational medical research encompassing advances in drug discovery, drug repurposing, genetic variant annotation, and the automation of specific biomedical research tasks such as data collection, gene function annotation, or literature mining (Yu, Beam, & Kohane, 2018; Ahuja, 2019).

Moreover, AI is –well-suited to handle repetitive work processes, manage large amounts of data, and can provide another layer of decision support to mitigate errors, allowing for the improvement of patient outcomes while reducing treatment costs (Frost & Sullivan, 2016; Accenture, 2017). More specifically, AI promises to find and use complex underlying relationships between the way humans work and how to care for them to improve care, discover new treatments, and advance scientific hypotheses even if we as humans do not understand those underlying relationships (Price & Nicholson, 2019).

In the context of healthcare, AI is poised to play an increasingly prominent role in medicine and healthcare because human biology is tremendously complex, and our tools for understanding it are limited (Price & Nicholson, 2019). Advances in computing power, learning algorithms, and the availability of large datasets (big data) sourced from medical records and wearable health monitors have proved to be useful in overcoming this significant shortcoming, and that is why AI is applied in several healthcare areas (Ahuja, 2019; Custers, 2006). Thanks to the processing of vast amounts of health data from electronic health records, AI could help soon diagnose diseases as accurate as experienced pediatricians (Liang et al., 2019), predict women at high risk of postpartum depression (Zhang et al., 2021), or give triage advice safer than that of human specialists (Razzaki et al., 2018). AI improves diagnostic accuracy and efficiency in provider workflow and clinical operations, facilitates better disease and therapeutic monitoring, and enhances procedure accuracy and overall patient outcomes (Kaul, Enslin & Gross, 2020). Many companies, including the tech-giant IBM, combine advanced technology solutions, including AI, blockchain, and data analytics, to support digital healthcare transformations. Great progress exists in the field of oncology, which is very much image-based diagnosis systems, an end to which machines are excellent. IBM offers medical imaging solutions for clinicians to deliver more consistent care and tools for researchers looking to conduct efficient clinical trials.[1]

To this end, AI applications collect and analyze patient data and present it to primary care physicians alongside insight into a patient's medical needs and support predictive models that can be ulteriorly used to diagnose diseases, predict therapeutic response, and potentially develop preventative medicine in the future (Amisha et al., 2019). As such, AI is currently already being put to use for medical purposes within the fields of chronic disease management and clinical decision-making (Bresnick, 2016), radiology (Bakkar et al., 2018; Wang et al., 2017), oncology (Houssami et al., 2017; Patel et al., 2018), pathology (Cruz-Roa et al., 2017; Yu et al., 2016; Wong & Yip, 2018; Capper et al., 2018), dermatology (Haenssle et al., 2018), ophthalmology (Gulshan et al., 2016), cardiology (Yan et al., 2019; Petrone, 2018), gastroenterology (Wang et al., 2018), surgery (Hashimoto et al., 2018), and mental health (Topol, 2019). This all comes to show how recent and expected advances in AI technologies may entail incredible and unprecedented progress for medicine and healthcare delivery, both in terms of quantity and quality, that could eventually help repair diagnoses errors and their very high consequences for society soon (Singh, Meyer, & Thomas, 2014).

Along with improved computer hardware and software programs, digitized medicine has become more readily available, and AI in medicine has started to proliferate (Bakkar et al., 2018; Kaul, Enslin & Gross, 2020). In this sense, AI in medicine can be divided into two subtypes: virtual and physical (Amisha et al., 2019).

The virtual part ranges from electronic health record systems to neural network-based guidance in treatment decisions. In contrast, the physical part deals with robots assisting in performing surgeries, intelligent prostheses for people with physical disabilities, and elderly care. As such, AI-enabled computer applications will help primary care physicians to identify better patients who require extra attention and provide personalized protocols for each individual. Examples of such technologies are smartwatches that are capable of detecting atrial fibrillation (Buhr, 2017), and smartphone exams with AI are being pursued for a variety of medical diagnostic purposes, including skin lesions and rashes, ear infections, migraine headaches, and retinal diseases, such as diabetic retinopathy and age-related macular degeneration (e.g., AiCure) (Levine & Brown, 2018). Simultaneously, on the administration side of healthcare, AI applications automate non-patient care activities, such as writing chart notes, prescribing medications, ordering tests, allowing healthcare providers to cut documentation time, and improving reporting quality (Ahuja, 2019).

Despite all the promises of AI technology, it has shown formidable obstacles and pitfalls in its adoption and implementation in the healthcare setting, especially when it pertains to validation and readiness for implementation in patient care (Topol, 2019). A recent example of this is IBM Watson Health's cancer AI algorithm. When fed with very limited input (actual data) from clinicians, the potential for significant harm to patients and medical malpractice by a flawed algorithm arises. This highlights already existing concerns about the dangers resulting from so-called 'black-box algorithms' – complicated algorithms whose internal mechanism is not understandable for humans, even those who design them. Indeed, many machine learning tools provide detailed information and verdicts without always accompanying justification. This information then supports ulterior decision-making processes that may affect the life of patients tremendously. This opaqueness has led to an increased demand for transparency and explainability in AI environments (Felzmann et al., 2020) (e.g., see the explicit requirements for transparency laid down in the European Union's General Data Protection Regulation, GDPR) before an algorithm can be used for patient care in practice. All in all, this stresses the need for systematic debugging, audit, extensive simulation, and validation, along with prospective scrutiny before the relevant AI algorithm is unleashed in clinical practice (Topol, 2019).

2.3 THE HEALTHCARE ROBOT ECOSYSTEM

Ecosystems have the ability to generate powerful forces that can reshape and disrupt industries (McKinsey & Company, 2019). As noted earlier, in healthcare, they have the potential to deliver a personalized and integrated experience to consumers, enhance provider productivity, engage formal and informal caregivers, and improve outcomes and affordability. Here, the global management consulting firm McKinsey and Company defines an *ecosystem* as 'a set of capabilities and services that integrate value chain participants (customers, suppliers, and platform and service providers) through a common commercial model and virtual data backbone (enabled by seamless data capture, management, and exchange) to create improved and efficient consumer and stakeholder experiences, and to solve significant pain points or inefficiencies' (McKinsey & Company, 2020).

Today, the primary goal of healthcare provision is preventing and effectively managing chronic conditions. However, as we have shown, productivity in healthcare is lagging other services industries as these goals shift. New technologies promise care that is available nearby or at home, supports continuous self and autonomous care, and reduces friction costs between supporting stakeholders. It should be no surprise that over the past decades of growth, changes, and regulation, our healthcare system has grown to be increasingly complicated. This has caused a disconnect among those who populate it, and the key players often have conflicting interests and goals that make it impossible for them to unite to serve each of our unique needs, preferences, and values. The introduction, further improvement, and rapid deployment of digital technologies within the healthcare domain have only increased this complexity and magnified this disconnect. This can be seen in several concrete cases. One is the case of introducing robots such as surgery robots, in which a traditional surgeon-to-patient procedure is now a surgeon to robot-cloud-AI to patients, changing how the roles and responsibilities are distributed among the practitioners, manufacturers, support staff, and patients (Fosch-Villaronga et al., 2021). Another is the use of electronic health records, which is the collection of patient and population health information stored in a digital format.

As we can see, the healthcare ecosystem is the network of stakeholders, processes, and materials necessary for the treatment of an ailment by way of medical intervention on a patient (de Vries & Rosenberg, 2016). The extensive list of robotics stakeholders in general used in society identified in the European project 'RoboLaw,' which includes producers and employers of robots, insurance companies, trade unions, user associations, professional users, and policymakers (Palmerini et al., 2014), to a certain extent, can also be identified in the field of healthcare robots (Fosch-Villaronga et al., 2021), although this field calls for a more specific approach because of the many parties involved and the healthcare setting's particular nature.

Within the field of healthcare robots, several stakeholders can be identified (see Table 2.2), and many different actors use healthcare robots within a healthcare setting: doctors, medical professionals, patients, family members, caregivers, healthcare providers, or even technology providers. All these stakeholders have similar goals, although they experience healthcare from different viewpoints. These experiences range from providing (medical) care and independence and preserving patients' dignity, to empowering those with special needs (Simshaw et al., 2016). A common and practical approach in mapping the healthcare robot ecosystem is to divide the stakeholders in healthcare robotics into primary, secondary, and tertiary stakeholders (Riek, 2017). Here, primary stakeholders refer to those stakeholders that use healthcare robots on a regular or even daily basis. Within the category of primary stakeholders, Riek (2017) identifies direct robot users (DRU), clinicians (CL), and

Table 2.2 Healthcare Robot Categories and Definitions

Main Category Stakeholder	Subcategory Stakeholder	Description
Primary stakeholders	Direct robot users (DRU) Clinicians (CL) Caregivers (CG)	Primary stakeholders use healthcare robotics on a regular or even daily basis.
Secondary stakeholders	Robot makers (RM) Environmental service workers (ESW) Health administrators (HA)	Secondary stakeholders are involved in using healthcare robotics, but will not directly use them themselves.
Tertiary stakeholders	Policy makers (PM) Insurers (IC) Advocacy groups (AG)	Tertiary stakeholders are those parties who have an interest in the use and deployment of healthcare robotics in society, although it is unlikely that they will use them directly.

Source: Based on Riek (2017).

caregivers (CG); secondary stakeholders include those stakeholders that are involved in using healthcare robotics without directly using them. Within the category of secondary stakeholders, she identifies the robot makers (RM), the environmental service workers (ESW), and the health administrators (HA); and tertiary stakeholders refer to those parties who have an interest in the use and deployment of healthcare robots, although it is unlikely that they will use them directly. Finally, within the category of tertiary stakeholders, Riek (2017) distinguishes between policy makers (PM), insurers (IC), and advocacy groups (AG).

With the above information serving as an important basis, in the following chapters, we deep-dive into each of the above categories of healthcare robots. For each of these categories, we map their ecosystem and state of the art, and explain the use of AI within this context – starting with surgery robots.

NOTE

1 See https://www.ibm.com/watson-health/solutions/cancer-research-treatment.

REFERENCES

Accenture. (2017). Artificial intelligence: Healthcare's new nervous system. Accenture Insight Driven Health. Retrieved from https://www.accenture.com/_acnmedia/PDF-49/Accenture-Health-Artificial-Intelligence.pdf.

Ahuja, A. S. (2019). The impact of artificial intelligence in medicine on the future role of the physician. *PeerJ*, 7, e7702.

Amisha, Malik, P., Pathania, M., & Rathaur, V. K. (2019). Overview of artificial intelligence in medicine. *Journal of Family Medicine and Primary Care*, 8(7), 2328–2331.

Bakkar, N., Kovalik, T., Lorenzini, I., Spangler, S., Lacoste, A., Sponaugle, K., … Bowser, R. (2018). Artificial intelligence in neurodegenerative disease research: Use of IBM Watson to identify additional RNA-binding proteins altered in amyotrophic lateral sclerosis. *Acta Neuropathologica*, 135(2), 227–247.

Bresnick, J. (2016). Big data, artificial intelligence, IoT may change healthcare in 2017. Retrieved from https://healthitanalytics.com/news/big-data-artificial-intelligence-iot-may-change-healthcare-in-2017.

Buhr, S. (2017). FDA clears AliveCor's Kardiaband as the first medical device accessory for the Apple Watch. In *TechCrunch*. https://techcrunch.com/2017/11/30/fda-clears-alivecors-kardiaband-as-the-first-medical-device-accessory-for-the-apple-watch/.

Capper, D., Jones, D. T., Sill, M., Hovestadt, V., Schrimpf, D., Sturm, D., … Pfister, S. M. (2018). DNA methylation-based classification of central nervous system tumours. *Nature*, 555(7697), 469–474.

Cresswell, K., Cunningham-Burley, S., & Sheikh, A. (2018). Health care robotics – qualitative exploration of key challenges and future directions. *Journal of Medical Internet Research*, 20(7), 1–11. [e10410]. https://doi.org/10.2196/10410.

Cruz-Roa, A. et al. (2017). Accurate and reproducible invasive breast cancer detection in whole-slide images: A deep learning approach for quantifying tumor extent. *Scientific Report*, 7, 46450.

Custers, B. H. M. (2006). The risks of epidemiological data mining, In H. Tavani (ed.) *Ethics, Computing and Genomics: Moral Controversies in Computational Genomics*. Boston: Jones and Bartlett Publishers, Inc.

de Vries, C. R., & Rosenberg, J. S. (2016). Global surgical ecosystems: A need for systems strengthening. *Annals of Global Health*, 82(4), 605–613. https://doi.org/10.1016/j.aogh.2016.09.011.

Dolic, Z., Castro, R., Moarcas, A. (2019). *Robots in Healthcare: A Solution or a Problem?*. Luxembourg: Study for the Committee on Environment, Public Health, and Food Safety, Policy Department for Economic, Scientific and Quality of Life Policies, European Parliament. Retrieved from https://www.europarl.europa.eu/RegData/etudes/IDAN/2019/638391/IPOL_IDA(2019)638391_EN.pdf.

Donabedian, A. (1988). The quality of care: How can it be assessed? *JAMA*, 260(12), 1743–1748.

European Foresight Monitoring Network, EFMN. (2008). Roadmap robotics for healthcare. Foresight brief no. 157. Retrieved from, http://www.foresight-platform.eu/wp-content/uploads/2011/02/EFMN-Brief-No.-157_Robotics-for-Healthcare.pdf.

Fosch-Villaronga, E. & Drukarch, H. (2021). *On Healthcare Robots. Concepts, Definitions, and Considerations for Healthcare Robot Governance*. ArXiv pre-print, 1–87, https://arxiv.org/abs/2106.03468.

Fosch-Villaronga, E., Khanna, P., Drukarch, H., & Custers, B.H.M. (2021). A human in the loop in surgery automation. *Nature Machine Intelligence*, 1–2. https://doi.org/10.1038/s42256-021-00349-4.

Fosch-Villaronga, E., & Millard, C. (2019). Cloud robotics law and regulation: Challenges in the governance of complex and dynamic cyber–physical ecosystems. *Robotics and Autonomous Systems*, 119, 77–91. https://doi.org/10.1016/j.robot.2019.06.003.

Felzmann, H., Fosch-Villaronga, E., Lutz, C., & Tamò-Larrieux, A. (2020). Towards transparency by design for artificial intelligence. *Science and Engineering Ethics*, 1–29.

Frost & Sullivan. (2016). Frost & Sullivan From $600 M to $6 billion, artificial intelligence systems poised for dramatic market expansion in healthcare. Retrieved from https://ww2.frost.com/news/press-releases/600-m-6-billion-artificial-intelligence-systems-poised-dramatic-market-expansion-healthcare/.

Garbuio, M., & Lin, N. (2019). Artificial intelligence as a growth engine for healthcare startups: Emerging business models. *California Management Review*, 61(2), 59–83.

Gulshan, V., Peng, L., Coram, M., Stumpe, M. C., Wu, D., Narayanaswamy, A., ... & Webster, D. R. (2016). Development and validation of a deep learning algorithm for detection of diabetic retinopathy in retinal fundus photographs. *Jama*, 316(22), 2402–2410. https://doi.org/ 10.1001/jama.2016.17216.

Haenssle, H. A. et al. (2018). Man against machine: Diagnostic performance of a deep learning convolutional neural network for dermoscopic melanoma recognition in comparison to 58 dermatologists. *Annals of Oncology*, 29, 1836–1842.

Hashimoto, D. A., Rosman, G., Rus, D., & Meireles, O. R. (2018). Artificial intelligence in surgery: Promises and perils. *Annals of Surgery*, 268(1), 70–76. https://doi.org/10.1097/SLA.0000000000002693.

Houssami, N., Lee, C. I., Buist, D., & Tao, D. (2017). Artificial intelligence for breast cancer screening: Opportunity or hype? *The Breast*, 36(2017): 31–33. https://doi.org/10.1016/j.breast.2017.09.003.

Kaul, V., Enslin, S., & Gross, S. A. (2020). The history of artificial intelligence in medicine. *Gastrointestinal Endoscopy*, 92(4), 807–812.

Kim J., Gu G. M., Heo P. (2016). Robotics for healthcare. In Jo, H., Jun, H. W., Shin, J., Lee, S. (Eds.) *Biomedical Engineering: Frontier Research and Converging Technologies. Biosystems & Biorobotics*, Cham: Springer, *vol.* 9, pp. 489–509.

Lee, D., & Yoon, S. N. (2021). Application of artificial intelligence-based technologies in the healthcare industry: Opportunities and challenges. *International Journal of Environmental Research and Public Health*, 18(1), 271. https://doi.org/10.3390/ijerph18010271.

Levine, B. & Brown, A. (2018). Onduo delivers diabetes clinic and coaching to your smartphone. In *Diatribe*. https://diatribe.org/onduo-delivers-diabetes-clinic-and-coaching-your-smartphone.

Liang, H., Tsui, B. Y., Ni, H., Valentim, C. C., Baxter, S. L., Liu, G., ... Xia, H. (2019). Evaluation and accurate diagnoses of pediatric diseases using artificial intelligence. *Nature Medicine*, 25(3), 433–438.

McKinsey & Company. (2019). How the best companies create value from their ecosystems. https://www.mckinsey.com/~/media/mckinsey/industries/financial%20services/our%20insights/how%20the%20best%20companies%20create%20value%20from%20their%20ecosystems/how-the-best-companies-create-value-from-their-ecosystems-final.pdf.

McKinsey & Company. (2020). The next wave of healthcare innovation: The evolution of ecosystems. How healthcare stakeholders can win within evolving healthcare ecosystems. https://www.mckinsey.com/~/media/mckinsey/industries/healthcare%20systems%20and%20services/our%20insights/the%20next%20wave%20of%20healthcare%20innovation/the-next-wave-of-healthcare-innovation-the-evolution-of-ecosystems-vf.pdf?shouldIndex=false.

Mataric, A., Okamura, A., & Christensen, H. (2008). *A Research Roadmap for Medical and Healthcare Robotics*. Arlington, VA. pp. 1–30. Retrieved from http://bdml.stanford.edu/twiki/pub/Haptics/HapticsLiterature/CCC-medical-healthcare-v7.pdf.

Palmerini, E., Azzarri, F., Battaglia, F., Bertolini, A., Carnevale, A., Carpaneto, J., … Warwick, K. (2014). RoboLaw Project. D6.2 Guidelines on regulating emerging robotic technologies in Europe: Robotics facing law and ethics. Retrieved from http://www.robolaw.eu/RoboLaw_files/documents/robolaw_d6.2_guidelinesregulatingrobotics_20140922.pdf.

Patel, N. M. et al. (2018). Enhancing next-generation sequencing-guided cancer care through cognitive computing. *Oncologist* 23, 179–185.

Petrone, J. (2018). FDA approves stroke-detecting AI software. *Nature Biotechnology*, 36, 290.

Poulsen, A., Fosch-Villaronga, E., & Burmeister, O. K. (2020). Cybersecurity, value sensing robots for LGBTIQ+ elderly, and the need for revised codes of conduct. *Australasian Journal of Information Systems*, 24. https://doi.org/10.3127/ajis.v24i0.2789.

Price, I. I., & Nicholson, W. (2019). Medical AI and contextual bias. *Harvard Journal of Law & Technology*, 33(1), 1–52.

Razzaki, S., Baker, A., Perov, Y., Middleton, K., Baxter, J., Mullarkey, D., … Johri, S. (2018). *A Comparative Study of Artificial Intelligence and Human Doctors for the Purpose of Triage and Diagnosis*. arXiv preprint arXiv:1806.10698.

Riek, L. D. (2016). Robotics technology in mental health care. In *Artificial Intelligence in Behavioral and Mental Health Care* (pp. 185–203). Academic Press. https://doi.org/10.1016/B978-0-12-420248-1.00008-8.

Riek, L. D. (2017). Healthcare robotics. *Communications of the ACM*, 60(11), 68–78. https://doi.org/10.1145/3127874.

Randolph, A. & Miller, M. D. (1994). Medical diagnostic decision support systems—past, present, and future: A threaded bibliography and brief commentary, *Journal of the American Medical Informatics Association*, 1(1), 8–27, https://doi.org/10.1136/jamia.1994.95236141.

Sætra, H. S. (2020). The foundations of a policy for the use of social robots in care. *Technology in Society*, 63, 101383. https://doi.org/10.1016/j.techsoc.2020.101383.

Simshaw, D., Terry, N., Hauser, K., & Cummings, M. (2016). Regulating healthcare robots: Maximizing opportunities while minimizing risks. *Richmond Journal of Law and Technology*, 22(3).

Singh, H., Meyer, A. N., & Thomas, E. J. (2014). The frequency of diagnostic errors in outpatient care: Estimations from three large observational studies involving US adult populations. *BMJ Quality & Safety*, 23(9), 727–731.

Topol, E. J. (2019). High-performance medicine: the convergence of human and artificial intelligence. *Nature Medicine*, 25, 44–56.

Wang, N., Pierson, E. A., Setubal, J. C., Xu, J., Levy, J. G., Zhang, Y., ... & Martins Jr, J. (2017). The Candidatus Liberibacter–host interface: insights into pathogenesis mechanisms and disease control. *Annual review of phytopathology*, 55, 451–482. https://doi.org/10.1146/annurev-phyto-080516-035513.

Wang, P. et al. (2018). Development and validation of a deep-learning algorithm for the detection of polyps during colonoscopy. *Nature Biomedical Engineering*, 2, 741–748.

Wong, D. & Yip, S. (2018). Machine learning classifies cancer. *Nature*, 555, 446–447.

Yan, Y., Zhang, J. W., Zang, G. Y., & Pu, J. (2019). The primary use of artificial intelligence in cardiovascular diseases: what kind of potential role does artificial intelligence play in future medicine?. *Journal of geriatric cardiology: JGC*, 16(8), 585.

Yu, K. H. et al. (2016). Predicting non–small cell lung cancer prognosis by fully automated microscopic pathology image features. *Nature Communication*, 7, 12474.

Yu, K. H., Beam, A. L., & Kohane, I. S. (2018). Artificial intelligence in healthcare. *Nature Biomedical Engineering*, 2(10), 719–731.

Zhang, Y., Wang, S., Hermann, A., Joly, R., & Pathak, J. (2021). Development and validation of a machine learning algorithm for predicting the risk of postpartum depression among pregnant women. *Journal of Affective Disorders*, 279, 1–8.

3

AI FOR SURGICAL ROBOTS

3.1 SURGICAL ROBOTS ECOSYSTEM

It is approximately four decades ago that the first robot was used in the theater for surgery, and the extent to which the surgical fraternity has embraced robotic surgery since then has been unparalleled. Since then, many advances have been made in the field, partly due to the plethora of benefits afforded by robotics that are simply absent in traditional surgical methods – stability, accuracy, integration with modern imaging technology, greater range of motion, telesurgery, in addition to multiple other benefits inherent to individual surgical specialties (Shah, Vyas & Vyas, 2014). However, to fully utilize the potential of surgical robots, including their potential capabilities related to AI, it is essential to understand the past to build toward the future, starting once again with definitions and terminology.

Surgical robots are service robots supporting surgeons during surgical procedures. Importantly though, robots used in surgery are not always necessarily considered surgical robots (Chinzei, 2019). For instance, medical devices within the definition of robots exist in current surgeries like robot-shaped actuated operating tables or robotized microscopes, which are usually not considered surgical robots. Instead, they fall within the category of robotic surgical instruments. As such, a robotic surgical instrument is 'an invasive device with an applied part, intended to be manipulated by

DOI: 10.1201/9781003201779-4

robotically assisted surgical equipment (RASE) to perform surgery tasks' (IEC 80601-2-77:2019).

Since the mid-1980s, when the first robotic-assisted surgical procedures took place, surgical robotics has evolved into a highly dynamic and rapidly growing field of application and research, enjoying increasing clinical attention worldwide (Faust, 2007; Bergeles & Yang, 2013; Lane, 2018). Initially introduced for a limited type of surgical procedures, nowadays, advances in ergonomics, computing power, hardware dexterity, safety, and ease of surgery allow for the rapid adoption and dissemination of new technologies for robotic-assisted surgical procedures. Some of the most common procedures in this arena are in the field of cardiology or ophthalmology, but an increasing amount of minimally invasive surgical operations, meaning operations that involve the insertion of a narrow laparoscopic device into the human body instead of having to open up the patient to that end, are on the rise (Sridhar et al., 2017).

Over the years, robotic surgery, or robotic-assisted surgery (RAS), has gained popularity within surgical practice due to the extensive benefits that they hold. Among other things, operating with robots increases the accuracy, precision, and dexterity of the doctor's hands. Surgical robots also can prove excellent at tremor corrections, scaled motion, and haptic corrective feedback, which is very handy for allowing perfect movement and for letting surgeons know when they are touching organs. This all allows surgeons to conduct surgeries with a lesser chance of damage to the patient's body, more successful surgeries, and less invasive procedures that grant shorter patient recovery time and hospital stays, less pain, blood loss, noticeable scars and discomfort, and less risk of complications following the procedure (Boyraz et al., 2019; Jaffray, 2005). Moreover, since robots are devoid of shortcomings such as fatigue or momentary lapses of attention, they can perform repeated and tedious surgeries, enabling at the same time, the performance of surgical procedures that were previously considered impossible (Fosch-Villaronga et al., 2021). For instance, RAS could help 'optimize the production, distribution, and use of the health workforce and infrastructure; allocate system resources more efficiently; streamline care pathways and supply chains' in low- and middle-income countries (Reddy et al., 2019). Still, the rapidly increasing demand resulting from the beneficial responses to their use and the consequent demonstration of their practical clinical potential leaves an extensive amount of room for further development and innovation. Consider, for instance, the tremendous benefits that AI could offer within this context. Combining AI control algorithms with the built-in advantages of surgical robots may serve surgical practice extremely well by overcoming technical errors and shortening operative times, improving access to body areas that are usually very hard to reach without necessarily opening up a person's whole body (Panesar et al., 2019). Moreover, AI for surgeons can also help reduce and eventually remove human error, which is one of the leading causes of surgical complications.

At the same time, we should not become oblivious to the potential dangers that may simultaneously arise, somewhat resembling a future in which science fiction becomes a reality. While RAS benefits abound, introducing a robot to the doctor-to-patient relationship inevitably changes how surgeries are performed. RAS has made it possible to extend the abilities and capabilities of surgeons, but it also presents new challenges. For instance, a revision of 14 years of data from the Food and Drug Administration (FDA) has shown that robotic surgeons can cause injury or death if they spontaneously power down mid-operation due to system errors or imaging problems (Alemzadeh et al., 2016). Broken or burnt robot pieces can fall into the patient, electric sparks may burn human tissue, and instruments may operate unintendedly, all of which may cause harm, including death (Alemzadeh et al., 2016). Moreover, as surgical robots' perception, decision-making power, and capacity to perform a task autonomously will increase thanks to AI, the surgeon's duties and oversight over the surgical procedure will inevitably change (Fosch-Villaronga et al., 2021). Moreover, other issues relating to cybersecurity and privacy will become more significant (Yang et al., 2017). Security vulnerabilities may allow unauthorized users to remotely access, control, and issue commands to robots, potentially causing harm to patients (FDA, 2020). Many examples highlight the genuine risks of exploiting the vulnerabilities of cyber-physical systems in general (Fosch-Villaronga & Mahler, 2021). For instance, in 2015, a Jeep Cherokee was switched off remotely by hackers while being driven by a journalist. In another example, the Stuxnet virus subtly changed the speeds at which the Iranian nuclear centrifuges spun, thereby damaging and even destroying the carefully calibrated machines (Holloway, 2015). These cybersecurity risks are also relevant within the context of healthcare robots powered by AI because systems that exert direct control over the world can cause harm in a way that humans cannot necessarily correct or oversee (Amodei et al., 2016). Healthcare robots interact with humans and, in the healthcare sector, users are often in a vulnerable position, which makes these risks more critical. It is easy to picture how cyberattacks could have lethal consequences in the context of surgical robots using AI. For example, a malicious attacker could disrupt the behavior of a tele-operated robot during surgery and even take over such a robot because of the unrestricted and uncontrollable nature of communication networks. In 2015, some researchers triggered a sort of 'denial of service attack' and succeeded in stopping the robot from being adequately reset, impeding the procedure altogether, which could be fatal (Emerging Technologies from the Arxiv, 2015).

Despite the high-tech world we are currently living in, a scene like this is not as far from reality as you might initially be thinking. Despite the widespread adoption for minimally invasive surgery (MIS), a non-negligible number of technical difficulties and complications are still experienced during surgical procedures performed by surgical robots. To prevent or, at least, reduce such preventable incidents in the future, advanced techniques in the design and operation of robotic surgical systems and enhanced mechanisms for adverse event reporting ought to be adopted (Alemzadeh et al., 2016). Here too, AI might very well offer unprecedented potential if applied with careful consideration and toward the direction of correcting common humanly made errors.

To fully understand the role of surgical robots within the complex surgical environment, and pinpoint the areas for improvement, it is necessary to get a good grip and understanding of the ecosystem in which these robots are situated. The robotic surgery ecosystem is a smaller ecosystem within the complex healthcare robots' ecosystem, comprising the surgeon, the nurses, and other staff members, who help the doctor during the surgical procedure, and the patients as the direct robot users (Fosch-Villaronga et al., 2021). Also, the hospital administration plays an important role within this ecosystem, researching and procuring reliable measurements of processes costs, quality, and efficiency. Importantly, within this context, while the use and role of robots affect and influence other stakeholders, some of them – among which are the surgeon and support staff – will remain integral to the surgical environments for many functions, such as selecting the process parameters or positioning the patient, which further stresses the essential role humans still have in robot-mediated surgeries (Fosch-Villaronga et al., 2021).

3.2 SURGICAL ROBOTS' STATE OF THE ART

Contemporary literature is rich in providing examples and applications of surgical robots and the current and predicted developments within surgical robotics (Bergeles & Yang, 2013). Still, the field is very much scattered, and a clear, concise definition of *surgical robots* and an understanding of their precise applications are still lacking. To fill in the gaps and lack of clarity currently experienced in the field of surgical robotics, in the following we provide a structured overview of and further elaboration on the main categories currently established within the field of surgical robots, their purpose, context of use, and main characteristics. For this purpose, we define surgical robots as 'service robots that support surgeons during surgical procedures allowing for more accurate and minimally invasive interventions' (Boyraz et al., 2019; Fosch-Villaronga & Drukarch, 2021).

We distinguish between the traditional types of surgical procedures in categorizing surgical robots, namely open surgery and closed surgery. Open surgery is the traditional form of surgery, which primarily refers to the highly invasive procedure of making an – often large – incision and cutting the skin and human tissue so that the surgeon has a full view of the structures and organs involved. Based on his medical assessment, the surgeon then can determine and perform the necessary surgical procedures. While open surgery is generally considered a safe and effective type of surgery, it causes longer hospital stays, longer recovery periods, larger scars, more pain, and higher risks of complications (e.g., bleeding and infections). On the contrary, closed surgery refers to the minimally invasive technique involved in surgery that allows surgeons to perform surgical procedures by providing them access to the patient's body either through the body's natural openings or through small incisions in the body, and is only suitable under particular conditions (e.g., when there is no particular urgency or when the human capabilities lack the necessary precision). During the last three decades, MIS has influenced the techniques used in almost the entire field of surgical medicine, mainly due to the fact that this form of surgery allows surgeons to use various techniques to operate with less damage to the patient's body than would be the case with open surgery. As a result, MIS is generally associated with less pain and discomfort, shorter hospital stays, quicker recovery times, smaller scars, and less risk of complications following the procedure (Jaffray, 2005). In the future, and given these advantages, it is likely that the technology will get better and more procedures will be performed through robot technology.

Even more so, the continuing developments in MIS have led to the replacement of conventional surgery with minimally invasive surgical procedures, and they have also prompted surgeons to reevaluate conventional approaches to surgery. Nowadays, many surgical techniques fall within the scope of MIS. RAS typically falls within the scope of MIS (Boyraz et al., 2019), the difference here being that instead of the surgeon manually operating instruments, they – as primary users of the robotic surgical systems – are supported or replaced by the power and precision of high-tech robotic systems. However, here too we should not be blinded by the promises MIS offers as the introduction of these new approaches has, in some respects, led to significant drawbacks. For instance, the introduction and rapid adoption of MIS have largely prolonged learning curves for surgeons, increased costs due to the (high) investments needed to acquire the necessary equipment and instruments and led to longer operating times (Fuchs, 2002). It is also not clear how these advancements will affect the education of future medical doctors: Should medical schools start offering tech-based training to equip future generations with these skills?

As we are heading toward the future, robotic systems are increasingly beginning to equal human specialists at precision surgical tasks. They may even outperform human surgeons in precision, control, efficiency, and accuracy in the near future, although it is still some time away before this applies to all surgical procedures. Increasingly autonomous robotic assistance levels allow intricate surgical feats to be performed without the surgeon worrying that their hands might slip or their grip falter (Svoboda, 2019). However, in surgeries that are very high volume, for the time being human surgeons are still much better than robots at weighing their experience to make complex surgical judgments and develop contextual understanding, especially when faced with unexpected situations and circumstances (Svoboda, 2019). It is precisely this power and accuracy that increasingly allows robotic systems to perform MIS, a type of surgery that, by its very nature, requires a high level of precision.

To illustrate the wide presence of robots in current surgical procedures, Table 3.1 provides some examples of surgical procedures currently performed with the help of a robot.

When compared to the surgical environment prior to the introduction of robotics and in the early days of its operation, this all indicates that the field of RAS is rich in development and innovation. Surgical robots are used in different medical areas, usually on a spectrum that ranges from surgical robots that are more generic in nature to highly specific surgical robots (see Figure 3.1). One of the growing areas of application is pediatric and aging population cardiac surgeries. Although entirely autonomous robotic-assisted cardiac

Table 3.1 Surgical Procedures Performed with a Robot (Fosch-Villaronga & Drukarch, 2021)

Cardiac surgery	Ocular surgery
Cosmetic surgery	Orthopedic surgery
Dental surgery	Otorhinolaryngology
Endocrine surgery	Plastic and reconstructive surgery
Endoscopic surgery	Thoracic surgery
Gastrointestinal surgery	Urology
Gynecology	Vascular surgery

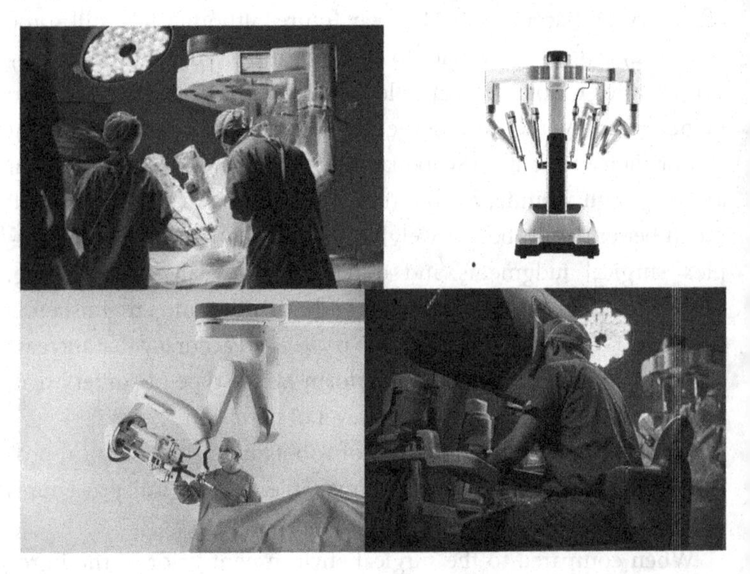

Figure 3.1 Intuitive Da Vinci robots used for RAS procedures.

surgeries are far from reality because these types of procedures are very long and complex, AI shows incredible promise in this arena. AI is a particularly relevant technology for cardiology, for instance, because the field is very much image-intensive (Chang, 2019). The techniques that involve AI, such as image processing and real-time automated decision-making processes, would help combat one of the largest and continuously climbing diseases among aging and pediatric populations.

3.3 THE APPLICATION OF AI IN SURGICAL ROBOTS

Owing to recent advances in medicine, AI has played an increasingly important role in supporting clinical decision-making and is now increasingly used for risk stratification, genomics, imaging and diagnosis, precision medicine, and drug discovery. AI was introduced into surgery more recently, gradually changing surgery practice with technological advancements in imaging, navigation, robotic intervention, and surgical instrumentation (Zhou et al., 2019). For instance, consider pre- and intra-operative imaging techniques such as ultrasound, computed tomography (CT), and magnetic resonance imaging (MRI). Moreover, advances in surgery have significantly impacted the management of both acute and chronic diseases, prolonging life and continuously extending the boundary of survival.

AI has the power to completely revolutionize contemporary surgery thanks to its combination with the processing of vast amounts of information and unlimited data storage, coupled with robotics, visualization, advanced instrumentation, data analytics, and connectivity (Aruni, Amit, & Dasgupta, 2018). In this sense, what AI adds to surgery robots is the potential use of decision-making algorithms to understand and react to specific data, making surgery performance more effective and reliable. In simple words: the more the data, the better the outcome. For example, a prostate recognition algorithm could make the machine learn whether a given image has prostate cancer or not, thus reducing the variability in radiologists' readings of magnetic resonance imaging. In another example, IBM's Watson created an intelligent surgical assistant that uses unlimited medical information, using natural language processing to clarify surgeons' doubts about surgery performance. Such a great outcome is because IBM Watson currently processes and analyzes an infinite total number of electronic medical records and sequence tumor genes to formulate more personalized and effective treatment plans. Moreover, with the increasing use of robotics in the surgical context, AI is set to transform the future of surgery through the development of more sophisticated sensorimotor functions with different levels of autonomy that can give the system the ability to adapt to constantly changing and patient-specific in vivo environments, leveraging the parallel advances in medicine in early detection and targeted therapy (Yang et al., 2017). The next-generation robots are associated with faster digital communication, better decision-making abilities, enhanced visual displays and guidance, and haptic feedback. With the development of increasingly sophisticated AI techniques, surgical robots can now achieve superhuman performance during MIS. Here, the objective of AI is to boost the capability of robotic surgical systems in perceiving the complex in vivo environment, conducting decision-making, and performing desired tasks with increased precision, safety, and efficiency. As a result, it is reasonable to expect that future surgical robots would perceive and understand complicated surroundings, conduct real-time decision-making, and perform desired tasks with increased precision, safety, and efficiency (Zhou et al., 2019). Because of this, it is hard to imagine that in the future, there will not be but more RAS procedures.

As established by Zhou et al. (2019), common AI techniques used for RAS can be summarized in the following four aspects: (1) perception, (2) localization and mapping, (3) system modeling and control, and (4) human–robot interaction. Depending on the AI capabilities and level of autonomy, surgical robots may be used for surgical procedures, ranging from less complicated surgeries on rigid body parts – think of the human bone complex – to more complex surgeries on soft human tissue – think of skin and organs (Prabu, Narmadha & Jeyaprakash, 2014). In this respect, the technology incorporated, including AI, and the surgical robot's embodiment play an essential role in the performance of surgical procedures. In this regard, extensive research on the current state of the art of surgical robots shows that surgical robots' main characteristics include robotic arm(s) used to mimic and extend human movement, cutting instruments, cameras, and X-ray systems. In addition, they comprise surgeon consoles and probes, and mobile compartments and tools, and lately AI. In practice, robotic platforms for surgical procedures involve an interplay between the sophisticated automated platform, on the one hand, and the surgeon, along with his/her team, on the other (Alemzadeh et al., 2016). The outcome of such shared task performance essentially depends on how they can be attuned to one another (Fosch-Villaronga et al., 2021).

3.4 SURGICAL ROBOTS' AUTONOMY LEVELS

Generally, robotic surgical systems operate within three different functions areas of medical practice, namely: (1) acquisition and analysis of information, (2) division of surgical trajectories or plan of actions, and (3) execution of the surgery (Manzey et al., 2009). As such, current surgical robots used to assist a surgeon performing (specific functions of) surgical procedures have different degrees of autonomy, ranging from no autonomy to full autonomy, and passing by being under the control of or in cooperation with a trained practitioner (Fosch-Villaronga et al., 2021). Remarkably, unlike the automation levels for automobiles by the standard SAE J3016 established by the Society of Automotive Engineers (SAE), currently, there are no universal standards that define the levels of autonomy in surgical robots. Nonetheless, Yang et al. (2017) have proposed a five-layered model for medical robotics autonomy levels, which has been further extended and refined in the literature (Varma & Eldridge, 2006; Yang et al., 2017; Ficuciello et al., 2019), and which has recently been further concretized by Fosch-Villaronga et al. (2021) as depicted in Figure 3.2.

Recent developments in RAS have suggested a strong affinity toward increased autonomy levels amongst stakeholders, with level 4 autonomy surgical robots currently being in the development stage. Based on the robots' capability and the surgeon's role in performing the desired task, surgical robots can, generally, be classified into three categories: (1) shared-controlled, (2) tele-controlled, and

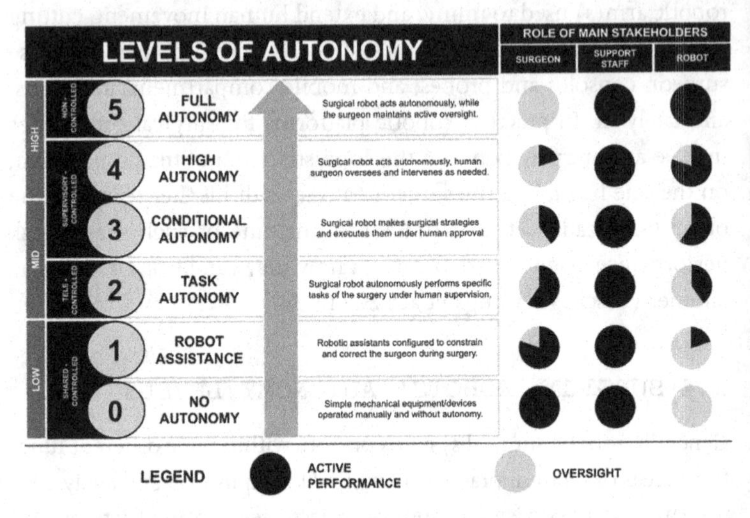

Figure 3.2 Autonomy levels and the role of humans in robot surgeries (Fosch-Villaronga et al., 2021).

(3) supervisory-controlled (Boyraz et al., 2019). The difference in these approaches held by surgical robots assisting in RAS primarily revolves around the robot's autonomy level, the degree of assistance provided by robotic systems during the execution of surgical procedures, and the human surgeon's control exercised (Fosch-Villaronga & Drukarch, 2021). To gain grip over each of these concepts of autonomy within the field of surgical robotics, we delve deeper into each of these below.

The **shared-controlled approach** refers to a surgical environment in which one or more robotic devices work together with the surgeon, meaning that the surgeon carries out the procedure with the use of a robot that offers steady-hand manipulations of the instrument, enabling the surgeon and the robotic system to perform the surgical procedure in question together (Mohammad, 2013; Fosch-Villaronga & Drukarch, 2021). The shared-controlled approach follows the method by which the workspace is divided into several segments and relates to the 0–1 levels of autonomy (see

Figure 3.2). Here, the robotic device behaves differently based on different localization – it responds to classifications such as safe, close, boundary, or forbidden – and is driven by haptic feedback (Boyraz et al., 2019) that will grow stronger as the surgeon's cutting tool comes closer to fragile human tissue, alerting the surgeon controlling the robot that extra caution should be taken.

The **tele-controlled approach** (also dubbed master-slave, remote-controlled, or telesurgical approach) allows a human surgeon to operate the robotic surgical device from a (close) distance with no pre-programmed or autonomous elements. This approach relates to the 2–3 levels of autonomy (see Figure 3.2). Already in 1996, the first FDA-approved robotic surgical system, ZEUS, was introduced. ZEUS is a complete robotic surgical system with seven degrees of freedom, tremor elimination and motion scaling (Ranev & Teixeira, 2020; Zemmar, Lozano, & Nelson, 2020), and was used for the first long-distance telesurgical procedure. Another breakthrough within this context was the da Vinci robotic system by intuitive surgery briefly introduced above, which is used across different surgical specialties for a variety of surgical procedures and is capable of performing technically challenging procedures (Marescaux et al., 2001; Troccaz, Dagnino & Yang, 2019; Zemmar et al., 2020). Generally, a tele-controlled surgical robotic platform consists of one or more robotic arms (slave element), to which the surgical instruments are attached using a console (dubbed a master controller), and which are generally configured with an optical system and computer-aided motion stabilization with a plurality of sensors for providing haptic feedback to the surgeon. As such, the telesurgical approach requires the surgeon to manipulate the robotic arms during the procedure rather than allowing the robotic arms to work according to a predetermined program (Mohammad, 2013) or directly and physically in tandem with the surgeon. Using real-time image feedback, the surgeon performing the surgical procedure can operate remotely using sensor data derived from the robot (Mohammad, 2013; Fosch-Villaronga & Drukarch, 2021). As such, the robotic arms form an extension of the surgeon's actual hands which is in line with the original intention of robotic surgery to permit the performance of a surgical procedure from a remote distance without touching the patient (Satava, 2002). Not surprisingly, since the outbreak of the COVID-19 pandemic, and the resulting unprecedented demands for hospitals, the tele-controlled approach has gained popularity (Fosch-Villaronga & Drukarch, 2021). To effectively reduce the spread of the virus in the hospital setting, robots and AI have increasingly been integrated into several sections of the surgical sequence, which each surgical patient traverses during a hospital stay to minimize contact between the patient and healthcare provider at each step (Zemmar et al., 2020). Digitization and machine intelligence have thus been called to action in the healthcare environment to combat the virus, and it is widely believed that their legacy may well outlast the pandemic and revolutionize surgical performance and management altogether (Zemmar et al., 2020).

The **supervisory-controlled approach** is the most automated of the three methods and relates to the 4–5 levels of automation (see Figure 3.2), and the robotic platforms following this approach generally comprise multiple robotic arms equipped with different surgical tools which are often powered by AI. The supervisory-controlled approach entails robotic systems configured to perform certain functions of the surgical procedure for a large part – although not yet fully – autonomously, with the surgeon being in a supervisory role throughout the procedure. The robotic devices are thus not yet capable of performing the concerning surgery without human guidance, as surgeons are often required to complete extensive preparations prior to the execution of the surgery and supervise during the execution thereof. As illustrated in Figure 3.2, the human surgeon's role changes from active performance to supervision. Likewise, the surgical robot's role transitions from supervision to active performance as the surgical robot's autonomy increases. However, while the surgeon thus remains indispensable in preparing and planning the surgical procedure and overseeing the execution thereof, he no longer partakes in the surgery's execution directly (Mohammad, 2013; Fosch-Villaronga et al., 2021).

In practice, this thus all means that surgical robots can perform certain functions of the surgical procedure independently. However, while Yang et al. (2017) state that the more autonomous medical robots become, the less human oversight exists, the phrasing of oversight may give the impression that RAS is gearing towards humanless surgeries, while this is not the case yet (see Figure 3.2) (Fosch-Villaronga et al., 2021). Research has pointed out that while the majority of RAS deploys surgical robots that follow the tele-controlled and shared-controlled approach, supervisory-controlled and fully autonomous surgical robotic devices have not yet found their way into RAS. Only performed on pigs so far, robot surgeons cannot perform an entire surgery completely independently from the beginning until the end on humans yet (Shademan et al., 2016; Greenemeier, 2020). Nevertheless, this does not withhold some researchers from conceptualizing them, and the efforts to move from the presently available level 3 robots toward level 4 robots indeed suggest that, in principle, the deployment of surgical robots with fully autonomous capabilities equivalent to level 5 is the ulterior motive of researchers and engineers working in this field (Yip & Das, 2017).

Still, surgery is not only about enhanced dexterity but also about context understanding. Until now, human surgeons have been shown to be considerably better than robots at weighing their experience to understand a particular context and make complex surgical judgments. In this sense, like cruise control and park assistance have made their way into cars progressively before realizing fully autonomous driving, fully autonomous surgical devices (level 5 of autonomy) will gradually enter clinical practice (Svoboda, 2019). Even though surgery robots operate increasingly autonomously, this thus does not mean that humans are completely out of the loop during surgical procedures. Humans will still perform many tasks and play an essential role in determining the robot's course of operation (Fosch-Villaronga et al., 2021). For instance, the medical support staff's role will remain integral and crucial to the surgical environments (e.g., for selecting the process parameters or for the positioning of the patient). Thus, even in the most autonomously performed surgeries, the medical staff will maintain an integral and crucial role within the surgical environment determining the robot's course of operation. Understanding who is responsible for what in this increasingly complex robotic ecosystem will especially prove to be of fundamental importance as surgical robots continue to be exposed to ever more sophisticated AI, allowing surgical robots to acquire new capabilities and achieve unprecedented levels of autonomy (Fosch-Villaronga et al., 2021).

REFERENCES

Alemzadeh, H., Raman, J., Leeson, N., Kalbarczyk, Z., Iyer, R.K. (2016). Adverse events in robotic surgery: A retrospective study of 14 years of FDA data. *PLoS One*, 11(4), e0151470. https://doi.org/10.1371/journal.pone.0151470.

Amodei, D., Olah, C., Steinhardt, J., Christiano, P., Schulman, J., & Mané, D. (2016). Concrete problems in AI safety. *arXiv preprint arXiv*:1606.06565. https://doi.org/10.48550/arXiv.1606.06565

Aruni, G., Amit, G., & Dasgupta, P. (2018). New surgical robots on the horizon and the potential role of artificial intelligence. *Investigative and Clinical Urology*, 59(4), 221.

Bergeles, C., & Yang, G. Z. (2013). From passive tool holders to microsurgeons: safer, smaller, smarter surgical robots. *IEEE Transactions on Biomedical Engineering*, 61(5), 1565–1576. https://doi.org/10.1109/TBME.2013.2293815.

Boyraz, P., Dobrev, I., Fischer, G., & Popovic, M. B. (2019). Robotic surgery. In Popovic, M. B. *Biomechatronics*, Academic Press, pp. 431–450.

Chang, A. C. (2019). Artificial intelligence in pediatric cardiology and cardiac surgery: Irrational hype or paradigm shift?. *Annals of Pediatric Cardiology*, 12(3), 191.

Chinzei, K. (2019). Safety of surgical robots and IEC 80601-2-77: the first international standard for surgical robots. *Acta Polytechnica Hungarica*, 16(8), 171–184.

Emerging Technologies from the Arxiv. (2015). Security experts hack teleoperated surgical robot. *MIT Technology Review*. Retrieved from https://www.technologyreview.com/2015/04/24/168339/security-experts-hack-teleoperated-surgical-robot/.

Faust, R. A. (2007). *Robotics in Surgery: History, Current and Future Applications*. Nova Publishers.

Ficuciello, F., Tamburrini, G., Arezzo, A., Villani, L., & Siciliano, B. (2019). Autonomy in surgical robots and its meaningful human control. *Paladyn, Journal of Behavioral Robotics*, 10(1), 30–43. https://doi.org/10.1515/pjbr-2019-0002.

Food and Drug Administration, FDA. (2020). Cybersecurity Safety Communications. Retrieved from https://www.fda.gov/medical-devices/digital-health/cybersecurity.

Fosch-Villaronga, E. & Drukarch, H. (2021). On Healthcare Robots. Concepts, Definitions, and Considerations for Policymaking. Available at ArXiv: https://arxiv.org/abs/2106.03468.

Fosch-Villaronga, E., Khanna, P., Drukarch, H., & Custers, B. H. M. (2021). A human in the loop in surgery automation. *Nature Machine Intelligence*, 1–1, https://doi.org/10.1038/s42256-021-00349-4.

Fosch-Villaronga, E., & Mahler, T. (2021). Cybersecurity, safety and robots: Strengthening the link between cybersecurity and safety in the context of care robots. *Computer Law & Security Review*, 41, 105528.

Fuchs, K. H. (2002). Minimally invasive surgery. *Endoscopy*, 34(02), 154–159. https://www.doi.org/10.1055/s-2002-19857.

Greenemeier, L. (2020). *Robot Surgeon Successfully Sews Pig Intestine*. Scientific American. Retrieved from https://www.scientificamerican.com/article/robot-surgeon-successfully-sews-pig-intestine/.

Holloway, S. (2015). Stuxnet Worm Attack on Iranian Nuclear Facilities. Online article. Retrieved from http://large.stanford.edu/courses/2015/ph241/holloway1/,

ISO. (2019). IEC 80601-2-77:2019; Medical electrical equipment – Part 2–77: Particular requirements for the basic safety and essential performance of robotically assisted surgical equipment. Retrieved from https://www.iso.org/standard/68473.html.

Jaffray, B. (2005). Minimally invasive surgery. *Archives of Disease in Childhood*, 90(5), 537–542.

Lane, T. (2018). A short history of robotic surgery. *Annals of the Royal College of Surgeons of England*, 100(6 sup): 5–7.

Manzey, D., Strauss, G., Trantakis, C., Lueth, T., Röttger, S., Bahner-Heyne, J. E., ... & Meixensberger, J. (2009). Automation in surgery: a systematic approach. *Surgical Technology International*, 18, 37–45.

Marescaux, J. et al. (2001). Transatlantic robot-assisted telesurgery. *Nature* 413, 379–380. https://doi.org/10.1038/35096636.

Mohammad, S. (2013). Robotic surgery. *Journal of Oral Biology and Craniofacial Research*, 3(1), 2. https://www.doi.org/10.1016/j.jobcr.2013.03.002.

Panesar, S., Cagle, Y., Chander, D., Morey, J., Fernandez-Miranda, J., & Kliot, M. (2019). Artificial intelligence and the future of surgical robotics. *Annals of Surgery*, 270(2), 223–226.

Prabu, A. J., Narmadha, J., & Jeyaprakash, K. (2014). Artificial intelligence robotically assisted brain surgery. *Artificial Intelligence*, 4(05).

Ranev, D. & Teixeira, J. (2020). History of Computer-Assisted Surgery. *Surgical Clinics of North America*, 100, 209–218. https://doi.org/10.1016/j.suc.2019.11.001.

Reddy, C. L., Mitra, S., Meara, J. G., Atun, R., & Afshar, S. (2019). Artificial Intelligence and its role in surgical care in low-income and middle-income countries. *The Lancet Digital Health*, 1(8), e384–e386. https://doi.org/10.1016/S2589-7500(19)30200-6.

Satava, R. M. (2002). Surgical robotics: the early chronicles: a personal historical perspective. *Surgical Laparoscopy, Endoscopy & Percutaneous Techniques*, 12, 6–16.

Shademan, A., Decker, R. S., Opfermann, J. D., Leonard, S., Krieger, A., & Kim, P. C. (2016). Supervised autonomous robotic soft tissue surgery. *Science Translational Medicine*, 8(337), 337ra64. https://www.doi.org/10.1126/scitranslmed.aad9398.

Shah, J., Vyas, A., & Vyas, D. (2014). The history of robotics in surgical specialties. *American Journal of Robotic Surgery*, 1(1), 12–20. https://doi.org/10.1166/ajrs.2014.1006.

Sridhar, A. N., Briggs, T. P., Kelly, J. D., & Nathan, S. (2017). Training in robotic surgery-an overview. *Current Urology Reports*, 18(8), 58.

Svoboda, E. (2019). Your robot surgeon will see you now. *Nature*, 573(7775), S110–S111. https://www.doi.org/10.1038/d41586-019-02874-0.

Troccaz, J., Dagnino, G. & Yang, G.-Z. (2019). Frontiers of medical robotics: from concept to systems to clinical translation. *Annual Review of Biomedical Engineering*, 21, 193–218. https://doi.org/10.1146/annurev-bioeng-060418-052502.

Varma, T. R. K., & Eldridge, P. (2006). Use of the NeuroMate stereotactic robot in a frameless mode for functional neurosurgery. *The International Journal of Medical Robotics and Computer Assisted Surgery*, 2(2), 107–113. https://doi.org/10.1002/rcs.88.

Yang, G. Z., Cambias, J., Cleary, K., Daimler, E., Drake, J., Dupont, P. E., Hata, N., Kazanzides, P., Martel, S., Patel, R. V., Santos, V. J., & Taylor, R. H. (2017). Medical robotics-Regulatory, ethical, and legal considerations for increasing levels of autonomy. *Science Robotics*, 2(4), eaam8638.

Yip, M., & Das, N. (2017). Robot Autonomy for Surgery. Retrieved from http://arxiv.org/abs/1707.03080.

Zemmar, A., Lozano, A. M., & Nelson, B. J. (2020). The rise of robots in surgical environments during COVID-19. *Nature Machine Intelligence*, 2(10), 566–572. https://doi.org/10.1038/s42256-020-00238-2.

Zhou, X. Y., Guo, Y., Shen, M., & Yang, G. Z. (2019). *Artificial Intelligence in Surgery*. arXiv preprint arXiv:2001.00627. https://arxiv.org/pdf/2001.00627.pdf.

4

AI FOR SOCIALLY ASSISTIVE ROBOTS

4.1 SOCIALLY ASSISTIVE ROBOTS ECOSYSTEM

Social robots represent a shift toward highly interactive robots (also called *socially interactive robots*) that entail deeper human–robot interaction (HRI) than other types of robots thanks to the capacity to interact with users *socially* (Breazeal, Dautenhahn, & Kanda, 2016). Socially assistive robots (SAR) intensify such HRI process and communication by providing direct support to users through social cues.

Due to the broad definition of assistance – which is generally defined as 'the act of helping or assisting someone or the help supplied' (Merriam-Webster, 2021) – and the almost infinite scope of social interaction, unlike surgical robots, the development of SARs cannot be aligned with or limited to a single purpose (Hegel et al., 2009; Li, Cabibihan, & Tan, 2011; Aymerich-Franch & Ferrer, 2020). The industry and promising research in the field of healthcare robotics are not oriented toward the optimization of a single task. Instead, SARs navigate the numerous entanglements between the needs of patients, the translation of those needs into concrete assistance, and how robots can modulate such assistance via social interaction (Fosch-Villaronga & Drukarch, 2021). The introduction of new and more advanced technologies, such as AI, to the SAR environment, is likely to lead to an even more confusing yet fruitful development of

DOI: 10.1201/9781003201779-5

devices that flirt with the boundary between medical devices, toys, and products (Fosch-Villaronga, 2019). SARs are a representation of the intersection between assistive robotics and socially interactive robotics. Here, assistive robots are service robots capable of assisting users (Tanaka et al., 2015), and Fosch-Villaronga and Drukarch (2021) define these as 'service robots assisting a user through physical or social interaction.' As such, SARs assist through social interaction (Feil-Seifer & Mataric, 2005).

4.2 SOCIALLY ASSISTIVE ROBOTS' STATE OF THE ART

The field of SARs is primarily oriented toward developing robots capable of close and effective social interactions to provide optimal assistance (Scassellati, Admoni, & Matarić, 2012), thereby acting in line with the concept of traditional caregiving. Unlike chatbots or other AI-driven assistive technologies, SARs specifically use their embodiment (arms, sensors, or touchscreens) to generate, modulate, and provide assistance through social interaction. This has resulted in a gradient spectrum of robot types that span from robots in which physical interaction has not a primary role, i.e., assistance is provided through social cues like the robot NAO; to a more complex mix of social/physical HRI, where robots use social cues and invite the user to have physical contact with them. The seal robot Paro is an example of a robot designed to interact with users socially (in a sort of human-animal relationship) and allow for physical contact with the patient, which has been proved to have enormous benefits for the patient (Fosch-Villaronga & Drukarch, 2021).

Typical embodiments for SARs include anthropomorphic, zoomorphic, caricatured, and functional (Fong, Nourbakhsh & Dautenhahn, 2003), and they play a significant and crucial role in many applications. For instance, children feel stronger friendship ties with a physically embodied robot compared to a virtual avatar, a physically present robot tutor produces better learning results, and individuals that suffer from cognitive impairments find the interaction with a physical robot more 'efficient, natural, and preferred' than with a simulated one (Tapus, Tapus & Mataric, 2009; Leyzberg et al., 2012; Sinoo et al., 2018). Moreover, robot embodiment increases presence, helps allocate social-interactional intelligence, typically via gaze and facial expressions, and makes robot task capabilities understandable from the user perspective (Tanaka, Nakanishi, & Ishiguro, 2015).

Due to the range of functions and applications falling under the umbrella of assistance, the various nomenclatures used to identify and point out these robots, and the fact that this subfield is still in the early stages of development, SAR categories blur, especially when compared to surgical robot categories. More so with the advancements in AI that empower these robots' capabilities incredibly.

To create more clarity around existing SAR categories, Libin and Libin (2005) have differentiated between social, educational, recreational, rehabilitation, and therapy robots, while other authors group the first three categories under the categories of 'care robots' (Vallor, 2011; van Wynsberghe, 2013). This indicates that the field clearly distinguishes between therapy, rehabilitation, and other care-related functions. To clarify the field, we distinguish SARs, building upon previous classifications, based on the type of assistance provided (Fosch-Villaronga & Drukarch, 2021), as depicted in Table 4.1.

We categorize SARs into therapy and care robots to provide a more explicit framework that accentuates the need for manufacturers

Table 4.1 Classification of Socially Assistive Robots (Fosch-Villaronga & Drukarch, 2021)

Category	Subcategory
Therapy	• Dementia • Autism • Neuro-developmental disorders
Care	• Companion1 • Basic assistance • Robot pet therapy2 • Aging-in-place in EU also called Active-Assisted Living[3] • Sex care robots

to state such robots' *intended use* to avoid misclassification and lack of the necessary safeguards to ensure safe use. Care robots primarily provide social interaction and support in any environment, including but not limited to a healthcare setting; for instance, think of a robot that keeps an elderly person company in their home. Therapy robots assist users with a specific form of therapy, which are condition and environment-specific, and thus used within controlled environments. The latter context of use implies that monitoring the interaction between the robot user and the robot is taking place. Although certain care robots are also used within a monitored environment, this distinction is unclear from the robot providers' perspective (Fosch-Villaronga & Drukarch, 2021). Therefore, it is necessary to obtain a clear understanding of the categories that make up SARs, what functions the different categories of SARs have, and how they are deployed in practice.

4.2.1 THERAPY ROBOTS

Therapy or therapeutic robots commonly cover robots used for *robotherapy*, a framework of HRI through which a series of coping skills are developed and mediated through robots (Libin & Libin, 2003). Here, the notion of *robotherapy*, like that of assistive robots, is, in practice, more oriented toward the physical (and more specifically, rehabilitation) than cognitive or psychological (Krebs & Hogan, 2006). Nevertheless, therapy robots also exist in socially assistive forms (Lorenz, Weiss & Hirche, 2016). They are used for existing therapies that serve a well-defined purpose (Rabbitt, Kazdin, & Scassellati, 2015) like socially assistive robotherapy, i.e., any form of psychological or cognitive therapy mediated through robots and, more specifically, through social robotic interaction (Libin & Libin, 2005).

4.2.2 CARE ROBOTS

The notion of care is vague and multiple, as is the case with the idea of assistance. While assistance may be a form of care, care can also be seen as a form of assistance. Care robots represent a prolific research domain in social robotics. Yet, care robots, assistive robots, and socially assistive robots are treated in the literature as distinct, independent categories without any relations of interdependence or entailment. An example of this separated treatment is in ISO 13482:2014, which goes even further by establishing the class of 'personal care robots' without clearly defining the meaning of personal care and excluding medical applications from its scope of application.

From the above categorization, it is clear that SARs state of the art illustrating the definitional opposition between the robot categories of care and therapy. Where one-on-one correspondences between the intended purpose and context of use are only visible for the robots falling under the therapy category, the very existence of care as a domain of application of healthcare technologies instead allows for the blurriness in the robots' application, notably revealing an unclear boundary between robots for serviceable contexts and healthcare robots (Fosch-Villaronga & Drukarch, 2021). This is notably seen with the robots NAO and Pepper, where healthcare is just another domain of application or vertical and not a field in its own right.

The application of AI within the subfield of SARs may further blur the understanding we have of the application and functions of SARs. It may simultaneously bring new and unimaginable areas of applications within the context of social assistance that have not been identified before and may entail significant progress for the field such as benefit SAR users. For instance, consider sex robots (Fosch-Villaronga & Poulsen, 2020). Sex robots are *service robots that perform actions contributing directly toward improvement in the satisfaction of the sexual needs of a user* (Fosch-Villaronga & Poulsen, 2020). Many authors believe that given their assistive capabilities, they could use sex robots within the context of SARs for disability care purposes. Generally, sex robots have different embodiments. These may include full or partial bodied humanoids, body parts such as arms, heads, or genitals used for sex-related tasks, or non-biomimetic robotic devices used for sexual pleasure.

Moreover, these robots usually display realistic sex-related body movements, have sensors to react in real time to user interaction, and include human cues such as voice, gaze, and lipsync to support human-like HRI. Interestingly, depending on their embodiment, sex care robots could also be classified as physically assistive robots. That is because besides satisfying sexual pleasures, these robots may help address first-time sex-related anxiety, treat sexual dysfunctions, or promote safer sex in educational settings. As such, sex robots could create a safe, non-judgmental environment for people who feel insecure about their sexual orientation (Levy, 2009; Royakkers & van Est, 2015). They may even be applied in more controversial contexts to treat pedophiles and potential sex offenders (Danaher, 2017) while also offering a means to meet the sexual needs of disabled and elderly individuals or as part of therapy for concerns such as erectile dysfunction, premature ejaculation, and anxiety surrounding sex (Sharkey et al., 2017).

4.3 THE APPLICATION OF AI IN SOCIALLY ASSISTIVE ROBOTS

SARs provide users with continuous support and personalized assistance through appropriate social interactions. Robots working in environments with people have to adapt to a constantly changing environment, which requires these robots to become more flexible by understanding human behavior and supporting users in heterogeneous tasks. This raises several challenges, including the need to realize intelligent and continuous behaviors, robustness and flexibility of services, and the ability to adapt to different contexts and needs (Umbrico et al., 2020). Here, AI plays a key role as it can realize cognitive capabilities like, for instance, learning, context reasoning, or planning that are highly needed in supporting real-time interaction, which is typical for socially assistive robots (Umbrico et al., 2020). More specifically, AI-powered cognitive technologies are designed to combine human intelligence with a range of AI capabilities such as machine learning, natural language processing, image analysis, and reasoning systems – creating an augmented intelligence that amplifies the impact of what humans and machines can do separately. When applied in the context of social assistance, such technologies can enable individuals in need to manage their well-being better and help strengthen and extend the social safety nets' reach to at-risk groups by addressing some of the critical challenges that typically impede provision and delivery, such as data inaccessibility, complexity, and the rate of caseworker churn (IBM, 2021).

Regarding therapeutic applications, research in embodied AI has indicated increasing clinical relevance, especially in mental health services – psychiatry, psychology, and psychotherapy (Fiske, Henningsen & Buyx, 2019). Within this field, technological applications range from 'virtual psychotherapists' to social robots in dementia care and autism disorder for elderly and children (see Figure 4.1) to robots for sexual disorders and mental disabilities.

Figure 4.1 LuxAI QTrobot used for therapy with children under autism spectrum disorder (ADS).

Moreover, increasingly, AI virtual and robotic agents are not mere deployed for low-level mental health support – think of comfort or social interaction –, but also for high-level therapeutic interventions that used to be offered exclusively by highly trained, skilled health professionals such as psychotherapists (Inkster, Sarda & Subramanian, 2018). In terms of AI-driven SARs used for therapeutic purposes, clinicians and scientists are increasingly exploring how innovations at the intersection of AI and robotics are translating into clinical practice (Fiske, Henningsen & Buyx, 2019). For instance, consider animal-like AI-driven companion robots such as Paro. Paro is a fuzzy haired seal, which is increasingly used for therapeutic purposes within the context of dementia because of its capability to engage individuals as at-home healthcare assistants, responding to speech and movement with dynamic dialog, or seeking to help elderly, isolated, or depressed patients through companionship and interaction. Research has already examined the role of such robots in reducing stress, loneliness, and agitation and improving mood and social connections with promising outcomes (Griffiths, 2014; Bemelmans et al., 2012).

Moreover, AI-driven SARs used for therapy have also shown great potential in engaging with children who have autism spectrum disorder (ASDs) (Scassellati, Admoni & Matarić, 2012). For instance, consider the Kaspar robot, which has demonstrated potential for integration in current education and therapy interventions (Huijnen et al., 2017) and is investigated to improve social skills among children with autism (Mengoni et al., 2017). Or consider the robot NAO, which has been designed to enhance facial recognition and appropriate gaze response, thereby teaching children who have trouble interacting with other people relevant social skills and the necessary means to apply these skills in practice in their relationship with human peers.

Finally, AI-driven SARs are also tested within other areas of mental health treatment such as mood and anxiety disorders, disruptive behavior (e.g., among children), and general assistance with mental health concerns (Rabbitt, Kazdin, & Scassellati, 2015).

A recent controversial example of such an application relates to the introduction of sex robots, whose raison d'être is simple:

Although every human should enjoy physical touch, intimacy, and sexual pleasure, persons with disabilities are often not in the position to fully experience the joys of life in the same manner as abled people (Fosch-Villaronga & Poulsen, 2020). Since the last decade, companies across the globe are increasingly offering adult sex robots, among which the famous Roxxxy, which are capable of speaking, learning their human partners' preferences, registering touch, and providing a form of intimate companionship thanks to the use of AI. AI-powered sex robots usually display realistic sex-related body movements, have sensors to support real time to user interaction, and include human cues such as voice, gaze, and lipsync to support human-like human–robot interactions.

All these advances reveal that the more AI technology advances, the more investigation will be needed around the benefits and opportunities AI has to offer in the areas of care and therapy. For example, would we be OK with having robots taking care of our grandparents or our children when ill? Are these tasks truly delegatable to machines? Since AI-powered SAR supports social HRI that happens more at the cognitive level than at the physical, special attention will have to be drawn to diversity and inclusion, tailor-made HRI responding to different personalities, special needs, and cultural backgrounds. Indeed, interacting socially with humans entails many aspects not apparent in other types of interaction. Think about sarcasm, sassy humor, or body language. Will robots be ready to handle human interaction thanks to AI? As automation also grows in care and therapy, society will be confronted with the question of whether these new avenues are desirable or not.

NOTES

1 Companion robots are also promoted as mental health robots as they lessen loneliness through the provision of robotic companionship.

2 Robot pet therapy is like robotherapy in the sense that it is defined simply as therapy with the medium of an animal. Most animal therapies are not diagnosis-specific but focus on alleviating side effects like loneliness.

3 Category coined by Lorenz, Weiss, and Hirche (2016). Comprise robots that facilitate care of the elderly (assist with tasks, remind when to take medicine etc.) and fall within the scope of healthcare services. *Also see* http://www.aal-europe.eu/.

REFERENCES

Aymerich-Franch, L., & Ferrer, I. (2020). The implementation of social robots during the COVID-19 pandemic. arXiv preprint arXiv:2007.03941, https://arxiv.org/abs/2007.03941.

Bemelmans, R., Gelderblom, G. J. , Jonker, P., de Witte, L. (2012). Socially assistive robots in elderly care: a systematic review into effects and effectiveness. *Journal of the American Medical Directors Association*, 13(2):114–120.e1. https://www.doi.org/10.1016/j.jamda.2010.10.002.

Breazeal C., Dautenhahn K., Kanda T. (2016). Social robotics. In: Siciliano B., Khatib O. (Eds.) *Springer Handbook of Robotics*. Springer, Cham, *Springer Handbooks*, 1935–1972.

Danaher, J. (2017). Robotic rape and robotic child sexual abuse: should they be criminalised?. *Criminal Law and Philosophy*, 11(1), 71–95.

Feil-Seifer, D., & Mataric, M. J. (2005). Defining socially assistive robotics. In *IEEE 9th International Conference on Rehabilitation Robotics*, ICORR 2005, 465–468. https://doi.org/10.1109/ICORR.2005.1501143.

Fiske, A., Henningsen, P., & Buyx, A. (2019). Your robot therapist will see you now: ethical implications of embodied artificial intelligence in psychiatry, psychology, and psychotherapy. *Journal of Medical Internet Research*, 21(5), e13216. https://doi.org/10.2196/13216.

Fong, T., Nourbakhsh, I., & Dautenhahn, K. (2003). A survey of socially interactive robots. *Robotics and Autonomous Systems*, 42(3–4), 143–166. https://doi.org/10.1016/S0921-8890(02)00372-X.

Fosch-Villaronga, E. (2019). *Robots, healthcare, and the law: Regulating automation in personal care*. Routledge.

Fosch-Villaronga, E. & Drukarch, H. (2021). *On Healthcare Robots. Concepts, Definitions, and Considerations for Healthcare Robot Governance*. ArXiv pre-print, 1–87, https://arxiv.org/abs/2106.03468.

Fosch-Villaronga, E., & Poulsen, A. (2020). Sex care robots: Exploring the potential use of sexual robot technologies for disabled and elder care.

Paladyn, Journal of Behavioral Robotics, 11(1), 1–18. https://doi.org/10.1515/pjbr-2020-0001.

Griffiths, A. (2014 July 08). How Paro the Robot Seal Is Being Used to Help UK Dementia Patients. *The Guardian*. URL: https://www.theguardian.com/society/2014/jul/08/paro-robot-seal-dementia-patients-nhs-japan.

Hegel, F., Muhl, C., Wrede, B., Hielscher-Fastabend, M., & Sagerer, G. (2009). Understanding social robots. In *IEEE 2009 Second International Conferences on Advances in Computer-Human Interactions*, 169–174.

Huijnen, C. A., Lexis, M. A., Jansens, R., de Witte, L.P. (2017). How to implement robots in interventions for children with autism? A co-creation study involving people with autism, parents and professionals. *Journal of Autism and Developmental Disorders*. 47(10), 3079–3096. https://www.doi.org/10.1007/s10803-017-3235-9.

IBM. (2021). Transforming social services. How cognitive technology is helping to protect the most vulnerable. https://www.ibm.com/watson/advantage-reports/ai-social-good-social-services.html.

Inkster B., Sarda S., Subramanian V. (2018). An empathy-driven, conversational artificial intelligence agent (Wysa) for digital mental well-being: real-world data evaluation mixed-methods study. *JMIR mHealth and uHealth*, 6(11), e12106. https://doi.org/10.2196/12106.

ISO. (2014). ISO 13482:2014; Robots and robotic devices - Safety requirements for personal care robots. Retrieved from https://www.iso.org/standard/53820.html.

Krebs, H. I., & Hogan, N. (2006). Therapeutic robotics: a technology push: stroke rehabilitation is being aided by robots that guide movement of shoulders and elbows, wrists, hands, arms and ankles to significantly improve recovery of patients. *Proceedings of the IEEE. Institute of Electrical and Electronics Engineers*, 94(9), 1727–1738. https://doi.org/10.1109/JPROC.2006.880721.

Leyzberg, D., Spaulding, S., Toneva, M., & Scassellati, B. (2012). The physical presence of a robot tutor increases cognitive learning gains. In *Proceedings of the Annual Meeting of the Cognitive Science Society*, 34(34), 1882–1887.

Levy, D. (2009). *Love and Sex with Robots: The Evolution of Human-Robot Relationships*. New York.

Li, H., Cabibihan, J. J., & Tan, Y. K. (2011). Towards an effective design of social robots. *International Journal of Social Robotics*, 3(4), 333–335.

Libin, E., & Libin, A. (2003). New diagnostic tool for robotic psychology and robotherapy studies. *CyberPsychology & Behavior*, 6(4), 369–374. https://doi.org/10.1089/109493103322278745

Libin, A., & Libin, E. (2005). Robots who care: robotic psychology and robotherapy approach. In *AAAI Fall Symposium: Caring Machines* (pp. 67–74). https://www.aaai.org/Papers/Symposia/Fall/2005/FS-05-02/FS05-02-011.pdf.

Lorenz, T., Weiss, A., & Hirche, S. (2016). Synchrony and reciprocity: Key mechanisms for social companion robots in therapy and care. *International Journal of Social Robotics*, 8(1), 125–143. https://doi.org/10.1007/s12369-015-0325-8.

Mehrotra, S., Kumar, S., Sudhir, P., Rao, G.N., Thirthalli, J., & Gandotra, A. (2017). Unguided mental health self-help apps: reflections on challenges through a clinician's lens. *Indian Journal of Psychological Medicine*, 39(5), 707–711. https://doi.org/10.4103/IJPSYM.IJPSYM_151_17.

Mengoni, S. E., Irvine, K., Thakur, D., Barton, G., Dautenhahn, K., Guldberg, K., et al. (2017). Feasibility study of a randomised controlled trial to investigate the effectiveness of using a humanoid robot to improve the social skills of children with autism spectrum disorder (Kaspar RCT): a study protocol. *BMJ Open*, 7(6), e017376. https://www.doi.org/10.1136/bmjopen-2017-017376.

Merriam-Webster (2021). *"Definition of Assistance"*. Merriam-Webster.com. https://www.merriam-webster.com/dictionary/assistance.

Rabbitt, S. M., Kazdin, A. E., Scassellati, B. (2015). Integrating socially assistive robotics into mental healthcare interventions: applications and recommendations for expanded use. *Clinical Psychology Review*. 35, 35–46. https://www.doi.org/10.1016/j.cpr.2014.07.001.

Royakkers, L., & van Est, R. (2015). *Just Ordinary Robots: Automation from Love to War*. CRC Press.

Scassellati, B., Admoni, H., & Matarić, M. (2012). Robots for use in autism research. *Annual Review of Biomedical Engineering*, 14, 275–294. https://doi.org/10.1146/annurev-bioeng-071811-150036.

Sharkey, N., van Wynsberghe, A., Robbins, S., Hancock, E. Foundation for Responsible Robotics. (2017). *Our Sexual Future with Robots*. URL: https://responsible-robotics-myxf6pn3xr.netdna-ssl.com/wp-content/uploads/2017/11/.

Sinoo, C., van Der Pal, S., Henkemans, O. A. B., Keizer, A., Bierman, B. P., Looije, R., & Neerincx, M. A. (2018). Friendship with a robot: Children's perception of similarity between a robot's physical and virtual embodiment that supports diabetes self-management. *Patient Education and Counseling*, 101(7), 1248–1255.

Tamura, T., Yonemitsu, S., Itoh, A., Oikawa, D., Kawakami, A., Higashi, Y., Fujimooto, T., Nakajima, K. (2004). Is an entertainment robot useful in the care of elderly people with severe dementia?. *The Journals of Gerontology: Series A*, 59(1), M83–M85, https://doi.org/10.1093/gerona/59.1.M83.

Tanaka, K., Nakanishi, H., & Ishiguro, H. (2015). Physical embodiment can produce robot operator's pseudo presence. *Frontiers in ICT*, 2, 8.

Tapus, A., Tapus, C., & Mataric, M. (2009). The role of physical embodiment of a therapist robot for individuals with cognitive impairments. In 18th *IEEE International Symposium on Robot and Human Interactive Communication, RO-MAN*, 103–107. https://doi.org/10.1109/ROMAN.2009.532621.

Umbrico, A., Cesta, A., Cortellessa, G. et al. (2020). A holistic approach to behavior adaptation for socially assistive robots. *International Journal of Social Robotics* 12, 617–637. https://doi.org/10.1007/s12369-019-00617-9.

Vallor, S. (2011). Carebots and caregivers: Sustaining the ethical ideal of care in the twenty-first century. *Philosophy & Technology*, 24(3), 251–268. https://doi.org/10.1007/s13347-011-0015-x.

van Wynsberghe, A. (2013). Designing robots for care: Care centered value-sensitive design. *Science and Engineering Ethics*. 19(2), 407–433. https://doi.org/10.1007/s11948-011-9343-6.

5

AI FOR PHYSICALLY ASSISTIVE ROBOTS

5.1 PHYSICALLY ASSISTIVE ROBOTS ECOSYSTEM

Over the past years, an increase in the demand for physical therapy services has been identified worldwide, and one of the reasons for this is the aging population. This resulted in the increased popularity of assistive technologies and rehabilitation robotics, especially as they promise to ease the stress on medical and physiotherapy staff and control expenses while simultaneously improving the lives of the physically, cognitively, or neurologically impaired (PCN-impaired) individuals (Fosch-Villaronga & Drukarch, 2021). Globally, many people suffer from various chronic physical, neurological, and cognitive disabilities (WHO, 2011). Seeing the significant advances in technology improve persons' independence and quality of life with these disabilities in many domains (Brose et al., 2010), it is for some years now that physically assistive robots (PAR) are increasingly deployed within the healthcare domain. Among their advantages include helping users walk back again, faster and more efficient rehabilitation with fewer resources. Moreover, they could eventually allow users to enter the workforce, lessen the burden on their caregivers, and live at home instead of in long-term care facilities, as medical complications are prevented, and self-image and life satisfaction are improved (Brose et al., 2010). As such,

DOI: 10.1201/9781003201779-6

physical assistance is one of the most direct ways that robots can help PCN-impaired persons (re)gain independence and function in physical tasks.

PAR can be defined as a 'personal care robot that physically assists a user to perform required tasks by providing supplementation or augmentation of personal capabilities' (ISO 13482:2014). In other words, PARs, or *exoskeletons*, are wearable robotic suits that support users in moving their arms and legs. In this sense, PARs may assist the robot operator, the person using the robot, or both, through physical interaction to perform specific tasks such as picking up a glass, opening a door, or walking around the house (Canal et al., 2017). As such, PARs interact with humans and can be directly worn by them as a sort of external skeleton (hence the word *exoskeleton*) that is on the user's body. Important to note within this context, however, is that this interaction is not limited to specific contexts. Thus, we can see exoskeletons being deployed in the industry (to help people bring heavy boxes from A to B) but also in the medical and rehabilitation context (to help them walk back again after a stroke). In this sense, the physical assistance provided by the PAR may be either partial, meaning that the robot acts as a supportive presence; or it may be total, meaning that the robot fully performs an action for the user (Fosch-Villaronga, 2019).

5.2 PHYSICALLY ASSISTIVE ROBOTS' STATE OF THE ART

PARs have a broad scope of use, and as a result, many different types of PARs can be identified, ranging from feeding robots to smart-powered wheelchairs and independent mobile robots to human–robot collaborative units (see Table 5.1). Generally, PARs can be divided into two specific categories: restraint and restraint-free PARs (ISO 13482:2014). While *restraint PARs* are fastened to the human body during use and directly assist PCN-impaired persons by being attached to them (usually for the lower or the upper limbs), *restraint-free PARs* are not fastened to the human body during use

Table 5.1 Classification of Physically Assistive Robots (Fosch-Villaronga & Drukarch, 2021)

Category	Subcategory
User support	• Exoskeletons and exosuits • Prosthetics • Robotic arms[1] • Walking aid (walkers, rollators) • Sensory-assistive robots (Hersh, 2015)
Task performance	• Feeding robots • Robotic manipulators • Smart wheelchairs • Robotic nursing assistants
User rehabilitation	• Orthoses[2] • End-effector robots • Exoskeletons[3]
Body part replacement	• Robotic prostheses

and, therefore, indirectly assist physically impaired persons (Fosch-Villaronga & Drukarch, 2021).

More specifically, a specific distinction can be made between supplementation and augmentation (ISO 13482:2014). Here, *supplementation* should be understood as the assistance that restores an average level of human capability to persons who may otherwise have difficulty doing so due to their disability, and *augmentation* refers to the physical aid in the performance of physical tasks that exceeds what can be generally expected without assistance (ISO 13482:2014; Fosch-Villaronga, 2019). When combined and in interaction with the user, these categorizations entail varying degrees of assistance, modulating the depth of the HRI and the user's capabilities that result from it (Fosch-Villaronga & Drukarch, 2021). These degrees reveal the increasing need to establish the different autonomy levels also for PAR. Although some efforts have been made in providing some specific-sector guidance on the autonomy levels for medical robotics (Fosch-Villaronga et al., 2021), more research is needed to understand such a complex intertwinement between the user and the device.

Fosch-Villaronga and Drukarch (2021) have established an overarching categorization for PARs based on the depth and complexity of the assistance provided by the PARs. More specifically, they distinguish between PARs that support the user(s) in the performance of specific tasks and within that, the distinction between upper or lower limb assistive, PARs that perform the physical task for its user(s), PARs used for rehabilitation, and PARs that function as body part replacements.

Before entering into detail within the different categories and what role artificial intelligence (AI) plays within the context of PAR, we want to highlight that Table 5.1 indicates that different assistive robots are developed and marketed to solve a specific issue and target a particular market. Still, it remains challenging to establish overarching categories that unify the different needs and problems solved by each sector's robot applications, in any case. While these categories may seem strictly separate from one another and other healthcare robot categories from the outset, in practice, they nevertheless appear to be of a more gradient nature, thereby often overlapping one another in their purpose and function (Fosch-Villaronga & Drukarch, 2021). For instance, consider robotic nurses. This mechanical system causes definitional blurriness in the healthcare domain, with some scholars seeing such robots as merely physical assistants and others indicating a strong desire for multipurpose and multifunctional robots, possibly leading to the existence of a mixed assistance category/trend heading in the direction of hybrid assistance (Hersh, 2015). This is because robotic nurses typically encompass a complex set of purposes and usages which are held by both PARs (e.g., if they can lift patients), socially assistive robots (e.g., if they have a social interface), and healthcare service robots (e.g., if they only bring medicines). Or consider exoskeletons, which can be deployed for both support and rehabilitation.

This illustrates that, contrary to the categorization of SARs, healthcare service robots (see Chapter 6), or surgical robots, PARs do not have robot applications that can be neatly distinguished between products and medical device categories. Instead, the same devices or types of devices are differentiated solely based on use, and then they can fall into these different categories at the same time. For instance, consider the robot ReeWalk used as a personal care robot and a stroke rehabilitation robot. This wearable robot would have a hard time finding the appropriate categorization within existing norms.

5.2.1 USER SUPPORT

Some exoskeletons support users in performing certain activities, for instance, in a warehouse. These exoskeletons focus on providing a supplementary physical force to the user's movement, so their work is easier, faster, and less heavy and in no case replaces the user in performing a task. To this end, these robots have been implemented in hospitals, for instance, in cases where nurses need to lift patients for activities of daily living. Instead of having a robot that lifts the patients for the nurses, the nurses are empowered with new robotic solutions that increase and augment their strength beyond what normal human capacity would be required. At the same time, such robots are used to support PCN-impaired individuals in the performance of specific tasks, which allows individuals who suffer from a severe physical or neurological disability that seriously limits their upper or lower limb mobility to perform ADL and vocational support tasks that would otherwise require a human attendant (Brose et al., 2010). For instance, consider robotic devices that control tremor correction and allow PCN-impaired individuals to feed themselves or grab objects in their surroundings. As such, the support offered by PARs can be considered more on the *augmentation* side of it or on the lower end of *supplementation*.

5.2.2 TASK PERFORMANCE

PARs used to support PCN-impaired individuals in the performance of specific tasks generally focus on applications that allow a person with a severe disability to perform ADL and vocational support tasks that would otherwise require a human attendant (Brose et al., 2010).

As such, the primary user of such PARs typically cover individuals who suffer from a severe physical or neurological disability that seriously limits their upper or lower limb mobility but can communicate clearly and have an average cognitive ability. Such disabilities are generally called into existence for people who suffer from high-level spinal cord injury (SCI), cerebral palsy, muscular dystrophy, and, more generally, for anyone who cannot manipulate household objects (Siciliano & Khatib, 2016). As such, PARs used for this purpose are usually under the control of a human operator, and their functionalities typically include handling books, medication, paper, computer media, food and drink, controlling communication devices, and activating electrical appliances. Examples of such robots that have been developed across the globe include feeding and drinking robots, bodyweight and body movement supportive robots, robotic arms, exoskeletons, prosthetics, and rehabilitation robots (see Figure 5.1).

PARs used for ADL task performance does not merely assist PCN-impaired individuals in a supportive sense but also fully perform the relevant task for the user. For instance, consider feeding robots (see Figure 5.2). These robots can promote independence and more intimacy during mealtimes (Herlant, 2018) and represent an end of the spectrum of robotic assistance where human control over the devices is indirect or, in the case of fully autonomous/automatic robots, non-existent (Fosch-Villaronga & Drukarch, 2021). Here, the degree of human–robot collaboration creates a range of autonomy in an action's performance, and within the context of PARs, this determines the degree of dependence the user has on the robot.

Moreover, it is important to stress that task performance robots assist all primary stakeholders, not just receivers of care. Due to the aging populations across the globe, there is a rising trend in robot nurses and the growing incorporation of robotic assistance into the healthcare ecosystem as part of the effort to optimize care systems by alleviating the routine tasks done by nurses. Such robots are a physical representation of human nurses and, as such, are capable of assisting doctors in the hospital context in the same way. We are already seeing

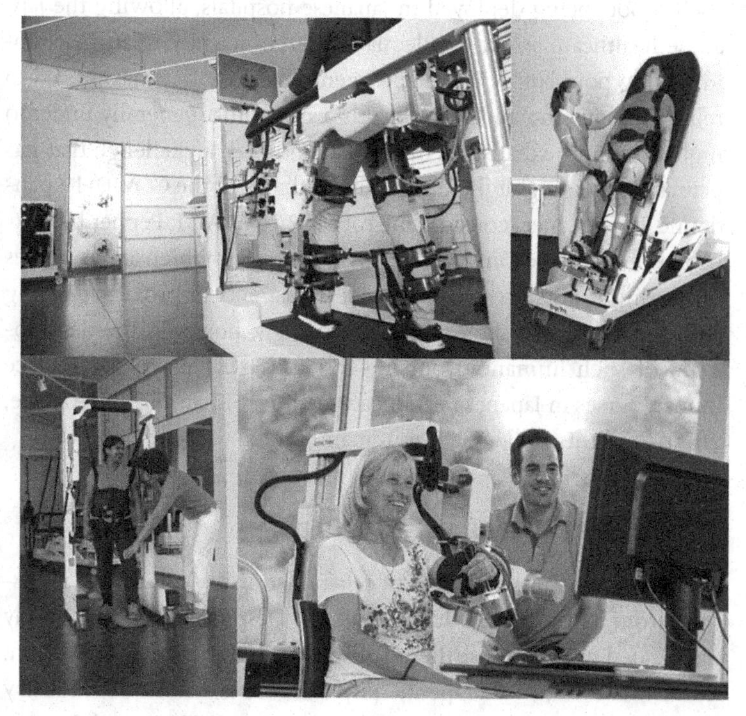

Figure 5.1 Rehabilitation robotics LokomatNanos, ErigoPro, Andago, and Armeo®Power (from top left to bottom right). (Pictures: Hocoma, Switzerland.)

Figure 5.2 The Obi feeding robot. Obi is a robotic feeding device for those with upper extremity limitations. Obi allows its users to choose what they eat and control the pace of their meal. Obi increases independence, social inclusion, and enjoyment at mealtime (www.MeetObi.com).

such robots being deployed in Japanese hospitals, allowing the Japanese healthcare sector to adequately and efficiently treat their rapidly aging population. The deployment of such robots has proved to relieve the nursing and other healthcare staff that generally undergo high stress and exhaustion due to patient load, a challenge that has primarily been highlighted since the outbreak of the COVID-19 pandemic. For instance, consider the robots Paro by AIST, Pepper by Softbank Robotics, and Dinsow by CT Asia Robotics, which are examples of nursing robots used to assist elderly patients in providing lifting and therapeutic assistance. Not surprisingly, however, the development of such humanoid nursing robots (HNR) may soon replace human nurses in Japanese healthcare facilities (Khan, Siddique & Lee, 2020), and this is likely not to remain limited to Japan.

5.2.3 USER REHABILITATION

PARs used within the context of rehabilitation are often seen as distinct from assistive robotics as a whole. Despite the growing body of blended applications, there still exists a definitional opposition, if not exclusion, of therapeutic physical robots within the industry (Fosch-Villaronga & Drukarch, 2021). Nevertheless, rehabilitation robots can still be considered PARs, which are applied in therapeutic settings, but like SAR, they are used in specific medical contexts under caregiver supervision and guidance. The assistance provided by PARs is generally limited in time and is determined based on the particular goal for which it is being deployed – think of the development of neuroplasticity required to regain lost motor functions (Gassert & Dietz, 2018). Moreover, rehabilitation robots can be further classified based on the different physical target areas they address (e.g., upper or lower limb), which again depends on the various degrees of cognitive and psychological engagement they require: grounded exoskeletons, grounded end-effectors, and wearable exoskeletons (Gassert & Dietz, 2018).

Generally, rehabilitation robots help patients recover from an accident or stroke throughout the rehabilitation process and assist and treat the disabled, elderly, and inconvenient people's conditions (Khan, Siddique & Lee, 2020). Moreover, rehabilitation robots promote functional reorganization compensation and regeneration of the nervous system, thereby effectively alleviating muscle atrophy (Zhao et al., 2020). At the same time, and as also established within the context of task performance PARs, rehabilitation robots permit rehabilitation physicians and staff room to relax from their overwhelming physical labor, thereby simultaneously optimizing the available healthcare resources. In practice, such robots include the Kinova assistive robot by Kinova Robotics, EksoNR by Eksobionics, and the wide range of robotics developed by Hocoma.

5.2.4 BODY PART REPLACEMENT

Body part replacement PARs typically cover medical implants, devices, or tissues placed inside or on the body's surface usually intended to replace missing body parts, such as a limb, a heart, or a breast implant, which may be lost due to physical injury, disease, or congenital conditions (FDA, 2021). Generally, the adjective robotic distinguishes passive adjustable devices from usually electrically powered mechatronic systems and stresses the presence of actuators, sensors, and microcontrollers and an intelligent control system implementing the desired behavior in these devices (Palmerini et al., 2014). Such medical implants can range from prostheses to devices that aid in delivering medication, monitoring body functions, or supporting organs and tissues, and while some of these implants are made from skin, bone, or other body tissues, others from metal, plastic, ceramic, or other materials. More specifically, these devices consist of three distinctive elements, namely (1) a biological (i.e., human or animal) part linked to (2) an artificial part (i.e., prosthesis, orthosis, or exoskeleton) using (3) a control interface (Micera et al., 2006). Significantly, prostheses differ from orthoses, as prostheses entail replacing a missing body part or an organ, while active orthoses improve the functionality of an existing body part (Palmerini et al., 2014). Moreover, these systems cover different levels of hybridness, augmentation, invasiveness, and temporality (Palmerini et al., 2014; FDA, 2021).

Here, hybridness refers to how close the artificial device and the human body are. These could be detached from the human body (i.e., tele-operated) or connected anatomically and functionally to the body, like a prosthesis. The augmentation concerns the number, type, and degree of human capabilities empowered, restored, or supported. Invasiveness refers to how invasive the biological and artificial elements connect, ranging from non-invasive (e.g., a joystick, or direct interfaces coupled with the central or peripheral nervous systems, such as brain–computer interfaces (BCI) (Micera et al., 2006)). In turn, these can be non-invasive, such as EEG, or invasive, such as implanted electrodes, which reveal another temporal dimension; that is, these systems can be placed permanently or removed once they are no longer needed.

PARs are typically characterized by various user interfaces and control systems that they incorporate and user perspectives that they generate. Here, user interfaces and control systems comprise the range of software and hardware components that allow a person with a disability to interact with their physically assistive robotic device (Brose et al., 2010). The operational modes of PARs can generally be classified as either shared-controlled (i.e., the user controls the system by continuously generating high-frequency motion commands and translating those from the control software into low-level functions) or supervisory-controlled (i.e., the user provides high-level low-frequency commands while the system operates entirely autonomously). From the user perspective, the operation of such devices may depend on the level of autonomy it requires (Arrichiello et al., 2017). These robotic devices' operation modes are strictly connected to the human-machine interface (HMI) used to generate and communicate commands (Arrichiello et al., 2017), and their ease of use depends on the level of workability of the user interface.

Importantly, these robotic devices are becoming more sophisticated thanks to the integration of AI. The advent of three-dimensional joysticks has made control of devices with more degrees of freedom possible (Gieschke et al., 2008). These mechanisms have already been extended to include chin and head interfaces, sip-and-puff (Fehr et al., 2000), voice control (Cagigas & Abascal, 2004), eye gaze direction (Yanco, 1998), EMG (Han et al., 2003), gesture- and intention-based human–robot interfaces, muscle-based robot interfaces, and BCIs. Notably, among the different HMIs, BCIs represent a relatively new technology that has recently been proposed to drive wheelchairs (Bi, Fan, & Liu, 2013; Carlson & Millan, 2013), guide robots for telepresence (Leeb et al., 2015; Escolano, Antelis, & Minguez, 2012), and control exoskeletons (Frisoli et al., 2012) and mobile robots (Gandhi et al., 2014; Riechmann, Finke, & Ritter, 2016). As control systems (continue to) become increasingly sophisticated and allow for larger 'bandwidths' of information to be transferred from human to machine, it is very likely that increasingly sophisticated devices will be developed.

5.3 THE APPLICATION OF AI IN PARs

The application of AI in medicine also promises significant progress for PARs. For instance, the increased capabilities with respect to advanced data acquisition, processing, and control techniques based on AI enable the construction of robust control strategies that outperform classic approaches in biomechatronic systems, including PAR (Vélez-Guerrero, Callejas-Cuervo, & Mazzoleni, 2021). AI can enable the development of increasingly sophisticated robots that can then take part in delivering care in a more efficient way by using some of the most popular techniques are based on different artificial neural networks (ANNs) and adaptive algorithms configurations, fuzzy logic, or other techniques to perform pattern detection or motion intention analysis (Cornet, 2013). These advances in AI and robotics can revolutionize the methods and capabilities of rehabilitation research and practice, enabling real-time interactions (Luxton & Riek, 2019), which are very much needed in PAR. Achieving real-time responses in PAR, especially lower-limb exoskeletons, is fundamental to achieve seamless integration with the user's (in this case, walking) movement.

Moreover, processing and control systems based on AI have progressively improved mobile robotic exoskeletons used in upper-limb motor rehabilitation (Cesta et al., 2018). More specifically, recent advancements in AI and its integration with robotics foster the diffusion of robotic agents with the personalized capabilities needed to support older adults and their caregivers in various situations and within a wide range of environments. AI-driven PARs can monitor and understand information coming from the environment in which they are placed, interacting with humans in a flexible and human-compliant way, autonomously performing tasks inside that environment, and personalizing interactions and services according to the specific needs of the user. In this sense, Cesta et al. (2018) stress that 'the ability to represent and reason diverse kinds of knowledge constitutes a key feature for allowing intelligent robotic assistants to understand the actual (and possibly time changing) needs of older persons as well as the status of the environment in which they are acting and inferring new knowledge to adapt their behaviors and better assist humans.' Consequently, many AI techniques must be integrated into a human–robot interaction loop to realize a needed set of advanced capabilities that completely match the needs of each individual user.

AI techniques constitute a fundamental enabler in realizing adaptive assistive services, such as those provided by PARs, to implement continuous monitoring and support in ADL, especially for the elderly. Such assistance is often offered in heterogeneous contexts and environments, which require such robotic systems to properly deal with to effectively support a person, and therefore, many features and capabilities must be taken into account (Cesta et al., 2018). In this sense, a set of critical requirements characterizing the capabilities of intelligent assistive robotic systems can be distinguished according to four correlated perspectives, namely: (1) environment perspective; (2) autonomy perspective; (3) interaction perspective; and (4) adaptation perspective.

- **Environment**. Several types of sensors can help gather information about a specific environment, including the internal context (the user's health status, the temperature of a room, the particular conditions of a room) but also the external context (if it is raining today if the hospital has an appointment with the person) (Fosch-Villaronga, 2019). The sensors are also divided between environmental or physiological sensors within the internal context (Cesta et al., 2018). For sure, the amount and kind of sensors in a specific context depend mainly on the particular purposes and objectives to be achieved by the project and what is legally permissible. Thanks to AI capabilities, the robots can deal with a constant flow of heterogeneous data coming from all of these sensors to monitor the state of the environment and autonomously recognize particular situations that require specific attention.
- **Autonomy**. Thanks to AI and the analysis of the information gathered from the environment, PARs can recognize particular situations that may require proactively executing supporting tasks. Think, for instance, if the robot detects that there is uneven terrain or there are stairs ahead and needs to adjust directly. This means that thanks to AI, robots will be able to have a safe and correct interaction of the system with the environment, understanding when they can autonomously decide the most optimized task sequence to achieve a particular objective. As Cesta and colleagues (2018) put it, a decision-making process is needed to achieve the level of autonomy needed to synthesize and carry out supportive actions automatically.
- **Interaction.** PARs usually interact with users in an active, physical manner (Fosch-Villaronga, 2019). To achieve seamless integration with the user's movement, PAR must correctly understand users' movement intentions (Tucker et al., 2015). Also, PAR must comply with social norms and what is socially expected from a specific movement. For instance, a feeding robot should not make abrupt movements in any case, at risk of embarrassing the user during mealtime.

- **Adaptation.** Everyone is different and has diverse habits and needs that can for sure change over time. In the specific context of PAR, a user may slowly recover the lost function or regain muscle strength after some robot usage. PARs should, after interacting with persons during their daily-home living or rehabilitation closely, adapt to the new situation after some time. In other words, learn over time. Thanks to AI, this is possible and more manageable than it was before. Eventually, AI-powered PAR can build distinct user profiles over time according to its experience and interaction with the user and personalize its behaviors to different persons accordingly using cloud services (aka cloud robotics, Fosch-Villaronga & Millard, 2019). This would reduce the cost of these devices over time and allow for high-speed integration of these personal movements to a shared knowledge database and the other way round (Fosch-Villaronga, 2019).

To sum up, AI has provided the basis for developing more reliable, flexible, and adaptable systems that can be truly wearable (Vélez-Guerrero, Callejas-Cuervo, & Mazzoleni, 2021). Having lightweight structures that adapt in real time to user needs allows for more usability, trust, and success in the implementation of PAR.

NOTES

1 Not to be confused with exoskeletons as these are robotic arms that attach to wheelchairs or tabletops and support the user's movements. Some models include slings within which patients can place their wrists or elbows. Other models offering performance-type assistance have been collected in the category robotic manipulators. See, for example, iFLOAT Arm.

2 Though robotic orthoses are often grouped under the label of exoskeletons, orthoses are medical devices to which Article 1.3 of the Medical Device Regulation (MDR) is applied as well as being defined in ISO 22523:2006.

3 Exoskeletons are included twice in this table as their use for medical applications is more regulated and context-specific than that of personal care (support).

REFERENCES

Arrichiello, F., Di Lillo, P., Di Vito, D., Antonelli, G., & Chiaverini, S. (2017). Assistive robot operated via P300-based brain computer interface. In 2017 *IEEE International Conference on Robotics and Automation (ICRA)* (pp. 6032–6037). IEEE. https://doi.org/10.1109/ICRA.2017.7989714.

Bi, L., Fan, X., & Liu, Y. (2013). Eeg-based brain-controlled mobile robots: a survey. *IEEE Transactions on Human-Machine Systems*, 43(2), 161–176. https://doi.org/0.1109/TSMCC.2012.2219046.

Brose, S. W., Weber, D. J., Salatin, B. A., Grindle, G. G., Wang, H., Vazquez, J. J., & Cooper, R. A. (2010). The role of assistive robotics in the lives of persons with disability. *American Journal of Physical Medicine & Rehabilitation*, 89(6), 509–521. https://doi.org/10.1097/PHM.0b013e3181cf569b.

Cagigas, D., & Abascal, J. (2004). Hierarchical path search with partial materialization of costs for a smart wheelchair. *Journal of Intelligent and Robotic Systems*, 39, 409–431. https://doi.org/10.1023/B:JINT.0000026090.00222.40.

Canal, G., Alenyà, G., & Torras, C. (2017). A taxonomy of preferences for physically assistive robots. In 2017 26th *IEEE International Symposium on Robot and Human Interactive Communication (RO-MAN)* (pp. 292–297). https://doi.org/10.1109/ROMAN.2017.8172316.

Carlson, T., & Millan, J. (2013). Brain-controlled wheelchairs: a robotic architecture. *IEEE Robotics and Automation Magazine*, 20(1): 65–73. https://doi.org/10.1109/MRA.2012.2229936.

Cesta, A., Cortellessa, G., Orlandini, A., & Umbrico, A. (2018). Towards flexible assistive robots using artificial intelligence. In *AI* AAL@ AI* IA* (pp. 3–15).

Cornet, G. (2013). Robot companions and ethics: a pragmatic approach of ethical design. *Journal International de Bioéthique*, 24(4), 49–58. https://doi.org/10.3917/jib.243.0049.

Escolano, C., Antelis, J., & Minguez, J. (2012). A telepresence mobile robot controlled with a noninvasive brain–computer interface. *IEEE Transactions on Systems, Man, and Cybernetics, Part B (Cybernetics)*, 42(3), 793–804. https://doi.org/10.1109/TSMCB.2011.2177968.

Fehr, L., Langbein, W. E., & Skaar, S. B. (2000). Adequacy of power wheelchair control interfaces for persons with severe disabilities: A clinical survey. *Journal of Rehabilitation Research and Development*, 37(3), 353–360.

Food and Drug Administration, FDA (2021). Implants and prosthetics. Retrieved from https://www.fda.gov/medical-devices/products-and-medical-procedures/implants-and-prosthetics.

Fosch-Villaronga, E. (2019). *Robots, Healthcare, and the Law: Regulating Automation in Personal Care*. Routledge. https://doi.org/10.4324/9780429021930.

Fosch-Villaronga, E. & Drukarch, H. (2021). On healthcare robots. *Concepts, Definitions, and Considerations for Healthcare Robot Governance*. ArXiv pre-print. 1–87, https://arxiv.org/abs/2106.03468.

Fosch-Villaronga, E., Khanna, P., Drukarch, H., & Custers, B. H. (2021). A human in the loop in surgery automation. *Nature Machine Intelligence*, 3(5), 368–369, https://doi.org/10.1038/s42256-021-00349-4.

Fosch-Villaronga, E., & Millard, C. (2019). Cloud robotics law and regulation: challenges in the governance of complex and dynamic cyber–physical ecosystems. *Robotics and Autonomous Systems*, 119, 77–91. https://doi.org/10.1016/j.robot.2019.06.003.

Frisoli, A., Loconsole, C., Leonardis, D., Banno, F., Barsotti, M., Chisari, C., & Bergamasco, M. (2012). A new gaze-bci-driven control of an upper limb exoskeleton for rehabilitation in real-world tasks. *IEEE Transactions on Systems, Man, and Cybernetics, Part C (Applications and Reviews)*, 42(6), 1169–1179. https://doi.org/10.1109/TSMCC.2012.2226444.

Gandhi, V., Prasad, G., Coyle, D., Behera, L., & McGinnity, T. (2014). Eegbased mobile robot control through an adaptive brain–robot interface. *IEEE Transactions on Systems, Man, and Cybernetics: Systems*, 44(9):1278–1285. https://doi.org/10.1109/TSMC.2014.2313317.

Gassert, R., & Dietz, V. (2018). Rehabilitation robots for the treatment of sensorimotor deficits: a neurophysiological perspective. *Journal of Neuroengineering and Rehabilitation*, 15(1), 1–15. https://doi.org/10.1186/s12984-018-0383-x.

Gieschke, P., Richter, J., Joos, J., et al. (2008). Four-degree-of freedom solid state mems joystick. *Presented at the 21st International IEEE Conference on MicroMechanical Electrical Systems, Tuscon, AZ* (pp. 86–89).

Han, J. S., Bien, Z. Z., Kim, D. J., Lee, H. E., & Kim, J. S. (2003). Human-machine interface for wheelchair control with EMG and its evaluation. In *Proceedings of the 25th Annual International Conference of the IEEE Engineering in Medicine and Biology Society* (IEEE Cat. No. 03CH37439) (Vol. 2, pp. 1602–1605). IEEE. https://doi.org/10.1109/IEMBS.2003.1279672.

Herlant, L. V. (2018). *Algorithms, Implementation, and Studies on Eating with a Shared Control Robot Arm. Doctoral Dissertation*, Robotics Institute, Carnegie Mellon University.

Hersh, M. (2015). Overcoming barriers and increasing independence – service robots for elderly and disabled people. *International Journal of Advanced Robotic Systems*. https://doi.org/10.5772/59230.

ISO. (2014). ISO 13482:2014; Robots and robotic devices – Safety requirements for personal care robots. Retrieved from https://www.iso.org/standard/53820.html.

Khan, Z. H., Siddique, A., & Lee, C. W. (2020). Robotics utilization for healthcare digitization in global COVID-19 management. *International Journal of Environmental Research and Public Health*, 17(11), 3819. https://doi.org/10.3390/ijerph17113819.

Leeb, R., Tonin, L., Rohm, M., Desideri, L., Carlson, T., and Millan, J. (2015). Towards independence: a bci telepresence robot for people with severe motor disabilities. *Proceedings of the IEEE*, 103(6), 969–982. https://doi.org/10.1109/JPROC.2015.2419736.

Luxton, D. D., & Riek, L. D. (2019). Artificial intelligence and robotics in rehabilitation. In L. A. Brenner, S. A. Reid-Arndt, T. R. Elliott, R. G. Frank, & B. Caplan (Eds.), Handbook of Rehabilitation Psychology (pp. 507–520). American Psychological Association. https://doi.org/10.1037/0000129-031.

Micera, S., Carrozza, M. C., Beccai, L., Vecchi, F., & Dario, P. (2006). Hybrid bionic systems for the replacement of hand function. *Proceedings of the IEEE*, 94(9), 1752–1762.

Palmerini, E., Azzarri, F., Battaglia, F., Bertolini, A., Carnevale, A., Carpaneto, J., … & Warwick, K. (2014). RoboLaw Project. D6.2 Guidelines on regulating emerging robotic technologies in Europe: robotics facing law and ethics. Retrieved, from http://www.robolaw.eu/RoboLaw_files/documents/robolaw_d6.2_guidelinesregulatingrobotics_20140922.pdf.

Riechmann, H., Finke, A., & Ritter, H. (2016). Using a cvep-based braincomputer interface to control a virtual agent. *IEEE Transactions on Neural Systems and Rehabilitation Engineering*, 24(6):692–699. https://doi.org/10.1109/TNSRE.2015.2490621.

Siciliano, B., & Khatib, O. (Eds.). (2016). Robotics and the Handbook. In: Siciliano, B., Khatib, O. (eds)*Springer Handbook of Robotics*. Springer, Cham. https://doi.org/10.1007/978-3-319-32552-1_1.

Tucker, M. R., Olivier, J., Pagel, A., Bleuler, H., Bouri, M., Lambercy, O., ... & Gassert, R. (2015). Control strategies for active lower extremity prosthetics and orthotics: a review. *Journal of neuroengineering and rehabilitation*, 12(1), 1. https://doi.org/10.1186/1743-0003-12-1.

Vélez-Guerrero, M. A., Callejas-Cuervo, M., & Mazzoleni, S. (2021). Artificial intelligence-based wearable robotic exoskeletons for upper limb rehabilitation: a review. *Sensors*, 21(6), 2146. https://doi.org/10.3390/s21062146.

World Health Organization (WHO). (2011). *World Report on Disability*. World Health Organization. Retrieved from https://www.who.int/teams/noncommunicable-diseases/disability-and-rehabilitation/world-report-on-disability.

Yanco, H. A. (1998). Wheelesley: A robotic wheelchair system: Indoor navigation and user interface. In Mittal, V. O., Yanco, H. A., Aronis, J., & Simpson, R. (Eds.). *Assistive Technology and Artificial Intelligence. Lecture Notes in Computer Science* (Vol. 1458). Springer, Berlin, Heidelberg. https://doi.org/10.1007/BFb0055983.

Zhao, P., Zi, B., Purwar, A., & An, N. (2020). Special issue on rehabilitation robots, devices, and methodologies. *Journal of Engineering and Science in Medical Diagnostics and Therapy*, 3(2). https://doi.org/10.1115/1.4046325.

6

AI FOR HEALTHCARE SERVICE ROBOTS

6.1 HEALTHCARE SERVICE ROBOTS ECOSYSTEM

Beyond the use of robots for direct assistance during medical procedures and other therapeutic applications, within the healthcare domain, robots are also used to facilitate the delivery and support of doctors and other medical staff's work in another way. Generally, these robots assist in the delivery of medication and supplies, enhance patient-doctor contact, and clean hospital facilities (Cepolina & Muscolo, 2014). While these robots may not fit into the typical picture of a healthcare robot, as is the case with the robots presented throughout the previous chapters, these robots perform vital tasks within this sector and have distinct characteristics from mere industrial robots. We call these robots healthcare service robots (HSR).

It is essential to note from the outset that while there is no commonly accepted definition for *service robots*, they are distinctive from industrial robots. Interestingly, as the task of defining service robots has continued to evolve, its meaning has become more and more blurred, primarily due to the crossover between the industry and service sectors. The Fraunhofer Institute for Manufacturing Engineering and Automation (Fraunhofer IPA) (1993) defined *service robots* as 'freely programmable kinematic devices that perform services

DOI: 10.1201/9781003201779-7

semi-or fully automatically.' As such, *services* were defined as 'tasks that do not contribute to the industrial manufacturing of goods but are the execution of useful work for humans and equipment' (Schraft, 1993). On its side, ISO defined *service robots* as 'a robot that performs useful tasks for humans or equipment, excluding industrial automation applications' (ISO 8373:2012). The International Federation of Robotics (IFR), on the other hand, emphasized the robot's autonomy in their definition by defining the term as 'technical devices that perform tasks useful to humans' well-being in a semi or fully autonomous way' (IFR, 2015a). While industrial robots often operate in controlled domains or domains that are hostile to humans, service robots commonly function alongside humans and in a reasonably uncontrolled environment (Mettler, Sprenger & Winter, 2017). To illustrate this, consider mobile robots and automated-guided vehicles (AGV) used in automation applications and new environments such as hospitals (Holland et al., 2021).

From a healthcare perspective, a *service robot* is viewed as 'any machinery in a clinical setting that can perform tasks, either partially or fully autonomously, to provide a useful service for healthcare delivery, including internal management,' e.g., delivering and transporting goods or cleaning floors (Garmann-Johnsen, Mettler & Sprenger, 2014; IFR, 2014; Mettler, Sprenger, & Winter, 2017). While HSRs have gained immense popularity over the past decade, the story of HSRs has nevertheless not always been successful (Stone et al., 2016). More than 30 years ago, the first service robot – HelpMate – was introduced into the healthcare domain to function as a courier robot in hospitals carrying around deliveries such as meals and medical records (Evans et al., 1989). Although the development of this type of robot was relatively simple and similar to other robots developed in the healthcare sector, their use has for long not been mainstream. With the high and rising costs in the healthcare sector, social pressure for lower prices, labor shortages, and an increasingly sick and aging population, however, the market for healthcare robots has proved to be a very promising avenue for investment (Simshaw et al., 2015). In fact, in 2018, it was predicted that the demand for professional service robots to support healthcare staff would reach 38 billion USD by 2022 (Müller, 2018) to lower the workload of healthcare staff and aid in complex tasks that need to be carried out (Taylor et al., 2016), indicating the revived potential for increased productivity and resource efficiency that HSRs can offer to the healthcare sector.

Thanks to their high levels of autonomy, HSRs are generally believed to make the delivery of care and hospital management more effective and quick, reduce labor costs for repetitive and often tedious tasks, and improve healthcare practitioners' work (Fosch-Villaronga, 2019). More specifically, HSRs can streamline routine tasks, reduce the physical demands on human workers, and ensure more consistent processes (Mettler, Sprenger, & Winter, 2017). Moreover, HSRs can keep track of inventory and place timely orders; help make sure supplies, equipment, and medication are where they are needed at the appropriate time; ensure that hospital rooms are sanitized and readied for incoming patients quickly; and offer excellent sanitary tools which are vital in care settings. While this means that HSRs can replace jobs or assist in the performance of tasks traditionally performed by human healthcare workers, many tasks performed by HSRs will likely still require some degree of human intervention. For instance, consider vehicles that assist in delivering food from the hospital kitchen to a specific hospital room. While such an HSR may relieve human healthcare workers from the tasks of food transport, nurses will still be required to give the delivered food to the patient. In this way, hospitals can ensure that nurses spend more time with their patients rather than pushing very heavy trolleys around the hospital's corridors.

Moreover, as is the case with SARs, developments in HSRs cannot be aligned or limited to a single purpose. The field of HSRs covers a wide range of applications, which is not merely limited to the medical field or the hospital setting. For instance, consider the delivery robot Relay used in hotels, hospitals, and public spaces or the VGo telepresence robots used in healthcare, education, and business. These *service robots* offer immense opportunities to streamline the logistics of any particular industry that requires 24/7 transport, cleaning, and disinfecting. In this respect, since healthcare is a remarkably sensitive domain of application, inserting robots in such contexts is not straightforward and still differs from any other field (Fosch-Villaronga, 2019).

Although advances in robotic technology have traditionally been found in manufacturing, mainly due to the need for stronger-than-human machines that could help build cars, ships, and other dangerous activities, in the service sector, especially in the healthcare sector, these robots perform different types of tasks in interaction with humans. Sometimes new opportunities arise in developing service robots that aid patients with illnesses, cognition challenges, and disabilities, but other times, these robots can help healthcare organizations and settings deliver care in a more efficient way (Fosch-Villaronga & Drukarch, 2021). In our contemporary context, the COVID-19 pandemic has further triggered the development of service robots in the healthcare sector to overcome the difficulties and hardships caused by this virus (Aymerich-Franch & Ferrer, 2020). More specifically, clinical care was the second most extensive set of uses for robotic devices throughout the pandemic, all with ground robots (Murphy, Gandudi & Adams, 2020), allowing for quick diagnosis of and acute healthcare provision to patients with the coronavirus while simultaneously protecting healthcare workers by enabling them to work remotely and cope with surges in demand.

6.2 HEALTHCARE SERVICE ROBOTS' STATE OF THE ART

The literature related to HSRs lacks specific categories, with most HSRs merely described based on their characteristics. A remarkable similarity between all HSRs can be identified, which relates to the fact that they have all been developed to make the hospital's daily processes more manageable and efficient (Fosch-Villaronga & Drukarch, 2021). In an attempt to further clarify the domain of HSRs, Fosch-Villaronga and Drukarch (2021) differentiate between those HSRs that completely take over medical staff tasks (such as cleaning the floors of the hospital), HSRs that support routine (non-medical) tasks (such as bringing the medicines from the pharmacy to the patients' rooms), and those that facilitate specific tasks (for instance robots that support tele-remote care). Moreover, they indicate that another critical characteristic of HSRs relates to their level of autonomy; while some HSRs function fully autonomously (e.g., the delivery robot TUG), others still require a certain degree of human intervention to operate, and again others function autonomously to a certain extent and assist, rather than replace, medical staff. Especially concerning the latter HSR type, several stakeholders (e.g., medical professionals and patients) are involved in the robot's use and functioning to realize its purpose. For instance, consider the InTouch Health telepresence robots. While these robots provide virtual care and make it possible for a medical professional to contact patients from a distance, the robot cannot fulfill its envisioned function without the stakeholders' involvement.

Building upon existing literature, Fosch-Villaronga and Drukarch (2021) provide a general categorization for HSRs, namely routine task robots, telepresence robots, disinfectant robots (and types within these categories), delivery robots, automated dispensing robots, remote inpatient care robots, remote outpatient care robots, infection prevention robots, and general cleaning robots (see Table 6.1). At the heart of this distinction lies the specificity of assisting or replacing medical staff, the robot's autonomy level, and their primary function are:

6.2.1 ROUTINE TASK ROBOTS

Routine task robots can generally be defined as 'autonomous and mobile robots that assist medical staff with daily routine tasks such as delivering food and medicine, carrying linens, pushing beds, or transferring lab specimens' (Fosch-Villaronga & Drukarch, 2021). These robots are typically designed to perform everyday tasks to relieve the pressure on the medical staff. However, although these robots are

Table 6.1 Classification of Healthcare Service Robots (Fosch-Villaronga & Drukarch, 2021)

Category	Subcategory
Routine task robots	• Delivery • Automated-guided vehicles • Serving robots • Mobile robots or platform • Drones • Automated dispensing • Healthcare administration
Telepresence robots	• Remote Inpatient Care (RIC) • Remote Outpatient Care (ROC)
Disinfectant robots	• Infection prevention • General cleaning

fully autonomous and mobile and, in some cases, can replace the medical staff as a whole, they are not necessarily anthropomorphic or social (Simshaw et al., 2015). Within the context of routine task robots, three subtypes can generally be distinguished, namely:

1. Routine task robots designed to deliver medical goods and to move around appliances in hospitals or other medical environments (delivery robots);
2. Routine task robots used in automated processes, such as medicine dispensing in (hospital) pharmacies (automated dispensing robots); and
3. Administrative robots used in healthcare management.

6.2.1.1 DELIVERY ROBOTS

An important and often underestimated aspect of healthcare is its underlying logistics. As many materials are transported in hospitals each day, such as medicine, medical supplies, laboratory samples, food, and linen, the healthcare sector is the ideal space for the deployment of delivery robots. Due to the need to reduce these logistic processes' operational costs and deal more efficiently with the increasing pressure on the supporting logistic functions and the rising demand for materials and equipment, a growing interest in logistics automation in hospitals has been identified. Consequently, an emerging application for autonomous navigating robots is hospital delivery.

Another type of delivery robot currently used in the medical sphere is serving robots. Serving robots carry out heavy-duty tasks in hospitals where pushing and pulling of material are required and are also deployed to supply food and beverages, dispense drugs, remove unclean laundry, deliver fresh bed linen, and transport regular and contaminated waste to/from various patients residing in hospital (Ozkil et al., 2009; Mettler, Sprenger & Winter, 2017). Since the 1950s, AGVs have optimized logistics in factories, hospitals, and homes (Hassan, 2006). AGV's are capable of transporting materials through wire guidance, inertial guidance, or laser guidance, which means that they depend to a certain extent on a predefined route and system. An example of this is the TransCar© robot, a self-guided delivery robot that can deliver medication, linens, and meals in the hospital environment. More recently, mobile robots are rising (Acosta Calderon, Mohan & Ng, 2015). For instance, consider HelpMate, the first mobile robot designed and developed to deliver pharmacy supplies and patient records between hospital departments and nursing stations (Evans, 1994). The main difference between these two types of serving robots is their dependence on an established infrastructure and level of autonomy.

Finally, another type of delivery robot currently used in the medical sphere is drones, traditionally used outside the healthcare environment. Nevertheless, (autonomous) drones are increasingly used as innovative tools for medical equipment delivery (e.g., medicine, defibrillators, blood samples, and vaccines), and they often use global positioning systems (GPS) and other sensors to navigate automated ground stations to deliver medications in remote locations that lack adequate roads. The use of healthcare delivery drones makes it possible to quickly and efficiently, which is vitally important in healthcare settings and deliver medical care to places where this would not usually be the case. For instance, consider Matternet drones used to deliver medicines after Haiti's earthquake in 2010, or the German Parcelcopter by DHL Parcel, which provides medications, materials, and blood samples (Scott & Scott, 2017).

6.2.1.2 AUTOMATED DISPENSING ROBOTS

For nearly everyone, taking the right medicine and the right amount is vital. However, due to the complexities within the medication-use process, errors are inevitable. At the end of the 20th century, automated medication dispensing systems were introduced to the healthcare domain and implemented in (hospital) pharmacies (see Figure 6.1). Their purpose is typically to minimize medication dispensing errors, save time, and secure the drug and administration process (Boyd & Chaffee, 2019), and since their introduction,

Figure 6.1 A robot hand arranging and storing drugs in a pharmacy storage room. (Pictures: Shutterstock. https://www.shutterstock.com/image-photo/germany-dortmund-2512-pharmacy-storage-room-1597763938.)

a moderate decrease in medication dispensing errors has been identified (Jones, Crane & Trussel, 1989). Besides preventing medication errors, robots and automated processes within the medication-use process also reduce costs. Moreover, following the outbreak of the COVID-19 pandemic, automated dispensing robots have gained significant interest within the healthcare sector, being the third-largest use of robots throughout this pandemic within the field of prescription and meal dispensing, whereby carts are navigated autonomously through a hospital (Murphy, Gandudi, & Adams, 2020).

6.2.1.3 HEALTHCARE ADMINISTRATION

Within the context of healthcare administration, robots are increasingly being deployed to streamline better routine administrative processes in hospitals and other clinical environments. Healthcare administrative robots are typically used at a hospital's reception to disseminate information about various units/sections of the hospital and guide patients and visitors. For instance, consider the robot Pepper and Dinsow 4 robot, which can handle several visitors without becoming tired and direct them to the physician of their choice. Moreover, they are exceptionally well received by children coming to the hospital, who experience their visit to the hospital as more pleasurable due to the interaction with the robot (Khan, Siddique & Lee, 2020). Interestingly, however, this field within healthcare currently seems to be directed toward using AI systems to boost healthcare administration rather than fully embodied robot technology (Fosch-Villaronga & Drukarch, 2021).

6.2.2 TELEPRESENCE ROBOTS

As we have already seen in Chapter 3, telepresence refers to a set of technologies used to create the impression that you are physically present in a remote place. As such, telepresence robots allow human operators to be virtually present and interact remotely through robot mobility and bidirectional live audio and video feeds

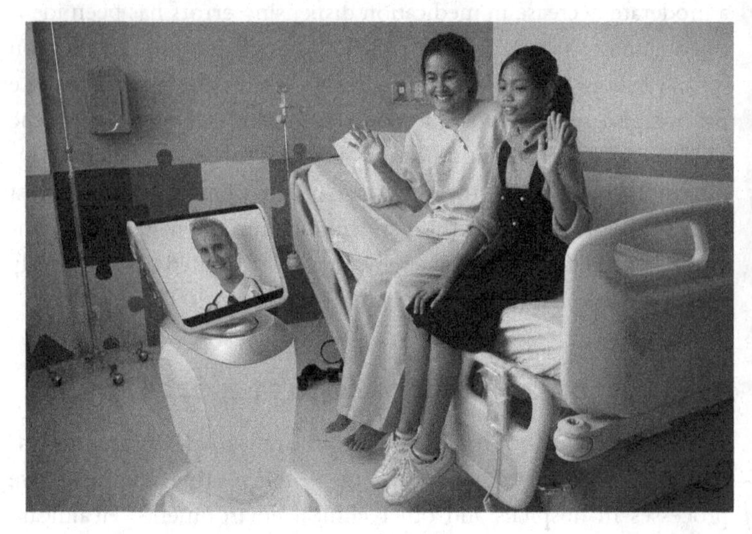

Figure 6.2 Doctor in the hospital talks with the patient in the patient room by telepresence robot and caretakers can interact with their patient to check on their living conditions and the need for further appointments. (Pictures: Shutterstock. https://www.shutterstock.com/image-photo/doctor-hospital-talk-patient-room-by-721490455.)

(Koceski & Koceska, 2016) (see Figure 6.2). Within the healthcare context, a growing interest has recently been expressed in developing telepresence robot systems for older adults' well-being (Koceski & Koceska, 2016), mainly as assistive technologies for telementoring in homes constitute a very promising avenue to decrease the load on the healthcare system, reduce hospitalization periods, and improve quality of life (Michaud et al., 2007).

Like other HSRs, telepresence robots' functions are not limited to healthcare but widely cover business and educational environments for videoconferencing or other commercial activities. In the healthcare context, telepresence robots proactively socially engage with users, creating an interaction with the person to give assistance and support in certain ADL and care (Broekens, Heerink & Rosendal, 2009; Feil-Seifer & Mataric, 2005). These robots can also help collect medical data about the vital signs of important patients for doctors and caregivers and are used for social interaction with other people (e.g., family members, friends, doctors, or caregivers). They can also help the elderly overcome a sense of social isolation and loneliness, affecting older people's physical, mental, and emotional health (Moren-Cross & Lin, 2006; Søraa et al., 2021). Generally, two types of telepresence robots can be distinguished within the context of the HSR category, based on the user, the remote environment where they function, and the stakeholders that interact with the robot, namely Remote Inpatient Care (RIC) and Remote Outpatient Care (ROC), with some telepresence robots covering both RIC and ROC. For telepresence robots used for RIC, the doctor is the primary user, using the robot within the hospital environment and in the interaction with nurses and patients. Here, the robot is mainly used for consultations and check-ups during hospitalizations. For instance, consider the physician-robot system, developed as a result of the InTouch Health Company and Johns Hopkins University's cooperation, which was one of the first telepresence robot systems allowing care assistance for the elderly and which enabled physicians to visit their hospitalized patients more frequently (Koceski & Koceska, 2016). Or consider the more recently developed RP-VITA by InTouch Health, which combines autonomous navigation and mobility, allowing doctors to monitor patients remotely (InTouch, 2011). For telepresence robots used for ROC, on the other hand, doctors remain the primary user but the remote environment in which the robot functions is the non-clinical environment (e.g., the patient's home and the robot now also interacts with the caregivers that surround the patients). Examples of such robots are the VGo robots by VGo Communications (Tsui & Yanco, 2013). While these robots are not different in terms of function and construction, they differ from RIC telepresence robots in the way they are implemented within the healthcare sector.

The ultimate goal of telepresence robots in healthcare is to provide specialized healthcare services over long distances. In this way, these robots can bridge the physical gap between the medical professional and the patient and make it possible to bring specialists and experts to remote areas where these services are currently unavailable (Kritzler, Murr, & Michahelles, 2016), such as third-world countries and war zones (Avgousti et al., 2016). Telepresence robots ensure a better quality of life in underdeveloped, isolated, or remote areas. Moreover, telepresence robots also significantly reduce the risks of transmitting infectious diseases among humans, which has become particularly important since the COVID-19 pandemic outbreak (Aymerich-Franch & Ferrer, 2020). The most prominent reported use of robotics within clinical care during the COVID-19 pandemic was for healthcare telepresence, including teleoperation by doctors and nurses to interact with patients for diagnosis and treatment (Murphy, Gandudi & Adams, 2020).

6.2.3 DISINFECTANT ROBOTS

Finally, while certainly not a new development within the healthcare environment, the COVID-19 pandemic has led to a surge in the development and adoption of disinfectant robots in the healthcare domain. Following the severe influx of patients and shortage of medical staff caused by the outbreak of the COVID-19 pandemic, and to reduce the exposure of medical staff to patients while simultaneously maintaining the social distancing guidelines, robots were rapidly deployed in hospitals and field hospitals to assist in cleaning and sterilizing (disinfecting) (Gupta et al., 2021). These robots typically allow healthcare workers on different levels to remotely monitor and manage their daily operations with robotic and autonomous solutions. The risk of exposure should remain low and restrict human interference to fewer subject areas to keep the risk of exposure low (Gupta et al., 2021). Generally, disinfectant robots can be divided into two subcategories, namely: (1) disinfectant robots used for general cleaning of hospital environments and other indoor and outdoor spaces; and (2) disinfectant robots used to sterilize work surfaces and medical equipment to prevent healthcare-associated infections (HAI) from infected patients (Khan, Siddique & Lee, 2020).

6.2.3.1 INFECTION PREVENTION

It is a long and well-known fact that the danger of new bacteria and infection by pathogens in healthcare environments, especially hospitals, is serious (Begić, 2017), and the consequences of HAIs include considerable pain, suffering, and even death (Begić, 2017), thereby constituting significant problems and high costs for the modern health sector. Advancements in robotic technology have triggered the usability of robotics in the next-generation healthcare system (Kaiser et al., 2021), which disinfect hospitals and other healthcare environments where they are believed to be of significant value in reducing the risk of hospital infections. Consequently, many new disinfection robots have been developed to help clean and disinfect these high-risk areas, including human support robots to sanitize high contact points and automated solutions for cleaning walls and floors. These robots are typically designed to emit a specific wavelength of ultraviolet light to the exposed surface to kill viruses and bacteria without exposing human personnel to infection (Guridi et al., 2019) and are generally remotely controlled from a safe distance. With sterilization methods not always being readily available and accessible, these robots offer a cost-effective solution to the manual disinfection of surfaces and objects, both in terms of time and minimization of the risk of exposure, to meet the remaining need for surfaces and things to be disinfected.

Moreover, within the context of disinfection in the healthcare environment, autonomous bots are often being transformed or adjusted to reassign tasks in the fight against novel coronavirus. For instance, the COVID-19 pandemic has accelerated the testing of robots and drones in public use, as officials seek out the most expedient and safe way to grapple with the outbreak and limit contamination and spread of the virus (Gupta et al., 2021). As such, for instance, drones designed initially to spray pesticides for agricultural applications have been quickly repurposed to spray disinfectants to fight against COVID-19 (Kaiser et al., 2021). An example of this is the multipurpose robot by the Chinese robot company, Youibot, which can monitor customers' temperatures using infrared cameras during the daytime and disinfect surfaces with the help of ultraviolet (UV) light in high-traffic areas, including hospitals, at night. Or consider, for instance, the Connor UVC Disinfection Robot, DJI, which is one of the companies that shared the responsibility to disinfect millions of square meters in China, and HAI by Xenex, which is widely adopted worldwide.

6.2.3.2 GENERAL CLEANING

Robots used for general cleaning within the healthcare environment are on the rise, especially since the outbreak of the COVID-19 pandemic. For instance, in China, robots have been assigned multiple tasks to minimize the spread of COVID-19 by utilizing them for cleaning and food preparation jobs in infected areas generally considered hazardous for humans (Khan, Siddique & Lee, 2020). These robots are typically used in hospital cleaning for dry vacuum and mopping (Prassler et al., 2000) and form an integral part of disinfecting hospitals to remove germs and pesticides. Examples include the Roomba cleaning robot by iRobot, an intelligent navigating vacuum pump for dry/wet mopping, UVD robot by UVD Robots ApS, a UV radiation-based device used to disinfect hospital premises from microbes, the Peanut robot used to clean washrooms of hospitals by using a highly dynamic robotic gripper and sensing system, and Swingobot 2000 by TASKI, a heavy-duty cleaning robot for cleaning hospital floors autonomously (Khan, Siddique & Lee, 2020).

The deployment of robots within the healthcare domain due to the outbreak of the COVID-19 pandemic is happening quicker than expected (Gupta et al., 2021). While technologies like telemedicine, telepresence, autonomous delivery robots, and sterilization robots have shown significant pragmatic promise well before the COVID-19 pandemic, they did not manage to achieve the success everyone hoped for various reasons (Gupta et al., 2021) until now. Due to the outbreak of COVID-19 world, many of these technologies have been fast-tracked to provide a near-to-normal lifestyle and protect citizens while simultaneously protecting healthcare workers from exposure to a contaminated environment and patients and alleviating the pressure on the healthcare sector worldwide. Moreover, beyond the clinical context, HSRs – especially telepresence technologies – are increasingly being deployed in another effort to reduce human contact and curb the spread of the virus through wearable devices and digital contact tracing apps used in various countries to identify individuals who have been in contact with an infected individual. These apps use Bluetooth and the user's geographical location, which is obtained via either the cellular network or an app installed on a smartphone to properly function, and are likely to include AI elements in the future.

6.3 THE APPLICATION OF AI FOR HEALTHCARE SERVICE ROBOTS

With the technology-based service encounter receiving significant attention following the advance of technology, especially AI-driven robots, HSRs are believed to be subjected to a wide range of innovations that are likely to dramatically change the healthcare service industry (Yoon & Lee, 2018; Wirtz et al., 2018). In more recent years, advancements in AI (e.g., development of deep neural networks, natural language processing, computer vision, and robotics) have brought into existence predictions that this new technology will take over health service activities currently being delivered by clinicians and administrators soon (Reddy, Fox, & Purohit, 2019). Even more so, it is precisely the likely impact of the infusion of robots in conjunction with AI and machine learning on frontline employees across the healthcare domain that has been attracting significant attention in recent years. While the abilities of service robots have gradually exceeded the performance capabilities of human service providers in areas related to memory, computing power, physical strength, and handling unpleasant or dangerous tasks, however, existing service robots are still characterized by a minimal level of AI integrated into their applications as part of their service provision capabilities (Chiang & Trimi, 2020). As such, the exceptional hype about the advanced abilities of AI within the healthcare service industry is inaccurate, especially the notion that AI will be able to replace human healthcare providers fully, but instead allows for the creation of a healthcare service system that could be termed an AI-enabled or AI-augmented (Reddy, Fox, & Purohit, 2019).

Generally, AI can be categorized as either mechanical, analytical, intuitive, or empathetic (Huang and Rust, 2018; Laowattana, 2020). Although the applications of service robots in use today have been subjected to a significant increase in demand due to the outbreak of the COVID-19 pandemic, they are typically used for the first two levels of AI as they still lack proficiency to reach the last two degrees of intelligence (Chiang & Trimi, 2020). As such, AI's intuitiveness and empathy intelligence have yet to be fully integrated into the capabilities of service robots to match or surpass that of humans (Huang & Rust, 2018). Consequently, it currently remains tough for service robots to independently perform complicated services in situations that require intuition, judgment, and empathy (Huang & Rust, 2018), features typically required within the healthcare domain. For precisely this reason, we can identify a gap between the level of service provided by AI-driven HSRs and their human counterparts, possibly even rendering HSRs useless within the context of specific service assignments (Chiang & Trimi, 2020). As a result, the HSR subfield is currently characterized by an augmentative relationship between robots and humans, as stressed above, where robots cannot replace human servers completely while simultaneously not being able to function without human assistance in providing quality service to patients (Baldwin, 2019). For instance, HSRs can clean and prep patient rooms independently, helping limit person-to-person contact in infectious disease wards, but are not yet capable of emotionally comforting patients throughout their hospital stay. At the same time, robots with AI-enabled medicine identifier software reduce the time it takes to identify, match, and distribute medicine to patients in hospitals, but they cannot motivate a patient to act per the doctor's prescription by convincing the patient of the necessity to do so.

Nevertheless, the potential roles of AI techniques in healthcare service delivery are becoming increasingly apparent, and studies have already highlighted the efficacy and potential of AI-enabled health applications (Agah, 2017; Ramesh, Kambhampati, & Drew, 2004). Especially within the context of healthcare administration, information technology tools have been demonstrated to alleviate the existing burden on health services, and AI and data mining techniques have been identified as among the most promising approaches to support healthcare administration as they are capable of augmenting clinical care and lessening administrative demands on clinicians (Snyder et al., 2011). For instance, AI-enabled HSRs can undertake repetitive and routine tasks like patient data entry and automated review of laboratory data and imaging results, and free time for clinicians to provide direct care. Moreover, HSRs can also use optimized machine learning algorithms to support clinic scheduling and patient prioritization, thus reducing waiting times and more efficient use of services. And to help hospitals in predicting the length of stay of patients at the pre-admission stage, thereby enabling more appropriate and efficient use of stretched hospital resources.

What we do not know yet is whether healthcare automation can also be reigned by the principles of resource efficiency and increased productivity as in other domains. The introduction of highly sophisticated machines in the healthcare domain may entail several changes, but the nature of these changes may not be immediately apparent if our focus is purely focused on the numbers. This is because current analyses of the changes caused by the insertion of a particular technology often fail to consider the broader consequences this may have at multiple levels, including the individual, the organizational, and the social. For instance, AI-powered HSR may have implications for new roles and responsibilities of medical practitioners and staff (individual changes), the allocation of responsibility and insurance (organizational change), or even the education of future health professionals (societal). Still, we cannot avoid the reality in which we live in, that is, that there is a continuous decline of human power willing to do the jobs that machines can do in a faster and more efficient way. At the end of the day, maybe AI will help us deliver better care although in ways we could not have imagined before.

REFERENCES

Agah, A. *Medical Applications of Artificial Intelligence*. Agah, A. (ed.). Boca Raton, FL: CRC Press, 2017, 461.

Avgousti, S., Christoforou, E. G., Panayides, A. S., Voskarides, S., Novales, C., Nouaille, L., Pattichis, C. S., & Vieyres, P. (2016). Medical telerobotic systems: current status and future trends. *BioMedical Engineering OnLine*, 15(96). https://doi.org/10.1186/s12938-016-0217-7.

Aymerich-Franch, L., & Ferrer, I. (2020). The implementation of social robots during the COVID-19 pandemic. arXiv preprint arXiv:2007.03941. https://arxiv.org/abs/2007.03941.

Baldwin, R. (2019). *The Globotics Upheaval: Globalization, Robotics, and the Future of Work*. Oxford University Press.

Begić, A. (2017). Application of service robots for disinfection in medical institutions. In *International Symposium on Innovative and Interdisciplinary Applications of Advanced Technologies* (pp. 1056–1065). Springer, Cham. https://doi.org/10.1007/978-3-319-71321-2_89.

Boyd, A. M., & Chaffee, B. W. (2019). Critical evaluation of pharmacy automation and robotic systems: a call to action. *Hospital Pharmacy*, 54(1), 4. https://doi.org/10.1177/0018578718786942.

Broekens, J., Heerink, M., & Rosendal, H. (2009). Assistive social robots in elderly care: a review. Gerontechnology, 8(2), 94–103. https://doi.org/10.4017/gt.2009.08.02.002.00.

Calderon, C. A. A., Mohan, E. R., & Ng, B. S. (2015). Development of a hospital mobile platform for logistics tasks. *Digital Communications and Networks*, 1(2), 102–111. https://doi.org/10.1016/j.dcan.2015.03.001.

Cepolina, F. E., & Muscolo, G. G. (2014). Design of a robot for hygienization of walls in hospital environments. In *ISR/Robotik 2014; 41st International Symposium on Robotics* (pp. 1–7). VDE.

Chiang, A. H., & Trimi, S. (2020). Impacts of service robots on service quality. *Service Business*, 14(3), 439–459. https://doi.org/10.1007/s11628-020-00423-8.

Evans, J., Krishnamurthy, B., Pong, W., Croston, R., Weiman, C., & Engelberger, G. (1989). HelpMate™: A robotic materials transport system. *Robotics and Autonomous Systems*, 5(3), 251–256. https://doi.org/10.1016/0921-8890(89)90049-3.

Evans, J. M. (1994). HelpMate: an autonomous mobile robot courier for hospitals. *Proceedings of IEEE/RSJ International Conference on Intelligent Robots and Systems (IROS'94), Munich, Germany, 1994*, (pp. 1695–1700, vol. 3). https://doi.org/10.1109/IROS.1994.407629.

Feil-Seifer, D., & Mataric, M. J. (2005). Defining socially assistive robotics. In *IEEE 9th International Conference on Rehabilitation Robotics* (ICORR 2005, pp. 465–468). https://doi.org/10.1109/ICORR.2005.1501143.

Fosch-Villaronga, E. (2019). *Robots, Healthcare, and the Law: Regulating Automation in Personal Care*. Routledge.

Fosch-Villaronga, E., Khanna, P., Drukarch, H., & Custers, B. H. M. (2021). A human in the loop in surgery automation. *Nature Machine Intelligence*, 1–1. https://doi.org/10.1038/s42256-021-00349-4.

Garmann-Johnsen, N., Mettler, T., & Sprenger, M. (2014). Service robotics in healthcare: a perspective for information systems researchers?. Retrieved from https://www.researchgate.net/profile/Tobias-Mettler/publication/267763443_Service_Robotics_in_Healthcare_A_Perspective_for_Information_Systems_Researchers/links/545a0a2f0cf2bccc4912fca5/Service-Robotics-in-Healthcare-A-Perspective-for-Information-Systems-Researchers.pdf.

Gupta, A., Singh, A., Bharadwaj, D., & Mondal, A. K. (2021). Humans and robots: a mutually inclusive relationship in a contagious world. *International Journal of Automation and Computing*, 1–19. https://doi.org/10.1007/s11633-020-1266-8.

Guridi, A., Sevillano, E., De La Fuente, I., Mateo, E., Eraso, E., & Quindós, G. (2019). Disinfectant activity of a portable ultraviolet C equipment. *International Journal of Environmental Research and Public Health*, 16(23) Article number 4747. https://doi.org/10.3390/ijerph16234747.

Hassan, Z. Z. (2006). *Automated Guided Vehicle (AGV) Using 68HC11 Microcontroller* (Doctoral dissertation, KUKTEM).

Holland, J., Kingston, L.; McCarthy, C., Armstrong, E., O'Dwyer, P., Merz, F., & McConnell, K. (2021). Service robots in the healthcare sector. *Robotics*, 10, 47. https://doi.org/10.3390/robotics10010047.

Huang, M. H., & Rust, R. T. (2018). Artificial intelligence in service. *Journal of Service Research*, 21(2), 155–172.https://doi.org/10.1177/1094670517752459.

International Federation of Robotics. (2015). *Definition of Service Robots*. Retrieved from http://www.ifr.org/service-robots/.

International Federation of Robotics. (2014). *Definition of Service Robots*. Retrieved from http://www.ifr.org/service-robots/.

InTouch. (2011). *InTouch Health Comprehensive Solutions*. Retrieved from http://www.intouchhealth.com/products.html.

ISO. (2012). ISO 8373:2012; Robots and robotic devices – Vocabulary. Retrieved from https://www.iso.org/standard/55890.html.

Jones, D. G., Crane, V. S., & Trussell, R. G. (1989). Automated medication dispensing: the ATC 212 system. *Hospital Pharmacy*, 24(8), 604–606.

Kaiser, M. S., Al Mamun, S., Mahmud, M., & Tania, M. H. (2021). Healthcare robots to combat COVID-19. In *COVID-19: Prediction, Decision-Making, and Its Impacts* (pp. 83–97). Springer, Singapore.

Khan, Z. H., Siddique, A., & Lee, C. W. (2020). Robotics utilization for healthcare digitization in global COVID-19 management. *International Journal of Environmental Research and Public Health*, 17(11), 3819. https://doi.org/10.3390/ijerph17113819.

Koceski, S., & Koceska, N. (2016). Evaluation of an assistive telepresence robot for elderly healthcare. *Journal of Medical Systems* 40, 121. https://doi.org/10.1007/s10916-016-0481-x.

Kritzler, M., Murr, M., & Michahelles, F. (2016). Remotebob: support of on-site workers via a telepresence remote expert system. In *Proceedings of the 6th International Conference on the Internet of Things*, 7–14.

Laowattana, D. (2020). Service robots head to pandemic's frontlines. Meet FACO: new breed of AI-powered, multi-tasking service robots. *Asian Robotic Review*. https://asianroboticsreview.com/home341-html.

Mettler, T., Sprenger, M., & Winter, R. (2017). Service robots in hospitals: new perspectives on niche evolution and technology affordances. *European Journal of Information Systems*, 26(5), 451–468. https://doi.org/10.1057/s41303-017-0046-1.

Michaud, F., Boissy, P., Labonte, D., Corriveau, H., Grant, A., Lauria, M. … & Royer, M. P. (2007). Telepresence robot for home care assistance. In *AAAI Spring Symposium: Multidisciplinary Collaboration for Socially Assistive Robotics*, 50–55.

Moren-Cross, J. L., and Lin, N. (2006). Social networks and health. In *Handbook of Aging and the Social Sciences* (6th ed.). New York: Elsevier. https://doi.org/10.1016/B978-012088388-2/50010-9.

Müller, C., (2019). *Market for Professional and Domestic Service Robots Booms in 2018*. Retrieved from https://ifr.org/post/market-for-professional-and-domestic-service-robots-booms-in–2018.

Murphy, R. R., Gandudi, V. B. M., & Adams, J. (2020). *Applications of robots for COVID-19 response*. arXiv preprint arXiv:2008.06976. https://arxiv.org/pdf/2008.06976.pdf.

Ozkil, A. G., Fan, Z., Dawids, S., Aanes, H., Kristensen, J. K., & Christensen, K. H. (2009). Service robots for hospitals: A case study of transportation tasks in a hospital. In *Proceedings of the 2009 IEEE International Conference on Automation and Logistics, Shenyang, China, 5–7 August* 2009. pp. 289–294.

Prassler, E., Ritter, A., Schaeffer, C. et al. (2000). A short history of cleaning robots. *Autonomous Robots*, 9, 211–226. https://doi.org/10.1023/A:1008974515925.

Ramesh, A. N., Kambhampati, C., Monson, J. R., & Drew, P. J. (2004). Artificial intelligence in medicine. *Annals of the Royal College of Surgeons of England*, 86(5), 334. https://doi.org/10.1308/147870804290.

Reddy, S., Fox, J., & Purohit, M. P. (2019). Artificial intelligence-enabled healthcare delivery. *Journal of the Royal Society of Medicine*, 112(1), 22–28. https://doi.org/10.1177/0141076818815510.

Scott, J., & Scott, C. (2017). Drone delivery models for healthcare. In *Proceedings of the 50th Hawaii International Conference on System Sciences*. https://doi.org/10.24251/HICSS.2017.399.

Schraft, R. (1993). Service robot – From vision to realization. *Technica*, 7, 27–31.

Simshaw, D., Terry, N., Hauser, K., & Cummings, M.L. (2015). Regulating healthcare robots: maximizing opportunities while minimizing risks. *Richmond Journal of Law & Technology*, 22, 1.

Snyder, C. F., Wu, A. W., Miller, R. S., Jensen, R. E., Bantug, E. T., & Wolff, A. C. (2011). The role of informatics in promoting patient-centered care. *Cancer Journal*, 17(4), 211. https://doi.org/10.1097/PPO.0b013e318225ff89.

Søraa, R. A., Nyvoll, P., Tøndel, G., Fosch-Villaronga, E., & Serrano, J. A. (2021). The social dimension of domesticating technology: Interactions between older adults, caregivers, and robots in the home. *Technological Forecasting and Social Change*, 167, 120678. https://doi.org/10.1016/j.techfore.2021.120678.

Stone, P., Brooks, R., Brynjolfsson, E., Calo, R., Etzioni, O., Hager, G. ... & Leyton-Brown, K. (2016). Artificial intelligence and life in 2030. One Hundred Year Study on Artificial Intelligence: Report of the 2015–2016 Study Panel, 52. Retrieved from https://ai100.sites.stanford.edu/sites/g/files/sbiybj9861/f/ai100report10032016fnl_singles.pdf.

Taylor, R.H.; Menciassi, A.; Fichtinger, G.; Fiorini, P.; Dario, P. (2016). Medical robotics and computer-integrated surgery. In *Springer Handbook of Robotics*. Springer: Berlin/Heidelberg, Germany. pp. 1657–1684.

Tsui, K. M. & Yanco, H. A. (2013). Design challenges and guidelines for social interaction using mobile telepresence robots. *Reviews of Human Factors and Ergonomics*, 9, 227–301. https://doi.org/10.1177/1557234X13502462.

Wirtz, J., Patterson, P. G., Kunz, W. H., Gruber, T., Lu, V. N., Paluch, S., & Martins, A. (2018). Brave new world: Service robots in the frontline. *Journal of Service Management*.

Yoon, S. N., & Lee, D. (2018). Artificial intelligence and robots in healthcare: What are the success factors for technology-based service encounters?. *International Journal of Healthcare Management*. https://doi.org/10.1080/20479700.2018.1498220.

CONCLUSION

Due to the demographic regression in developed countries, the number of persons who may potentially take care of older adults has dramatically decreased. For every person over 65 years of age, there are four people under that age capable of caring for that person. An aging population is condemned, however, to an inevitable and massive cost of healthcare which will be adversely impacted by the decline of welfare-state contributors if the birth rate continues to decline.

As such, technology can on many occasions be seen as a solution to the associated problems caused by aging, and there has been an increased number of technology applications in the field of care, including the delivery of certain services through the Internet (e-health), medical devices, mobile apps, and wearables.

Robotics and the application of AI within this context are transforming the healthcare domain. AI is becoming increasingly more sophisticated, and robots fueled by this new technology are becoming more capable of performing tasks that previously have typically been performed by humans, as they are capable of doing so more efficiently, quickly, and at a significantly lower cost. In the healthcare domain, AI is poised to play an essential role in helping prolong human life and allowing for more accessible care. While the benefits of the introduction of these new technologies to the healthcare domain are evident, questions may also be raised as to how and in what ways they could improve or hurt the quantity and quality of care (e.g., through more effective healthcare in which the process is arguably problematic) (Fosch-Villaronga & Drukarch, 2021).

This book is a stepping stone toward gaining a better understanding of AI in the context of healthcare robotics. Although usually put together within the rubric of 'healthcare robotics' or 'medical robotics,' robot surgeons, physical/socially assistive, and healthcare service robots largely differ in embodiment and context of use. Assistive technologies are meant to help patients, elderly, and disabled people in their daily needs, either in the hospital or at home. They may enhance a person's physical capabilities and may help nurses take care of patients better. In the context of surgery, surgical robots help perform minimally invasive, which is basically more beneficial for patients than traditional surgeries. Physically assistive robots also help users walk again or pick up a glass even if their condition may not allow them to do so.

As one can imagine, the use and development of robot technology that takes care of this particular part of the population raise many questions: Will robots be able to take care of humans in the future? Is healthcare automation a good thing? What are the activities that humans can (or should) delegate to machines? Will robotized healthcare become more personalized or on the contrary more alienating? Will robots deliver good care? This is even more certain with respect to the increased levels of autonomy of healthcare robots, in which their complex interaction with humans will inevitably blur practitioners' and developers' roles and responsibilities and affect society.

AI for healthcare robotics promises an unparalleled potential for healthcare providers, patients, and society. However, healthcare robots' full deployment will soon require more clarity on the division of responsibilities channeling robot autonomy and human performance, support, and oversight. Having a clear framework determining the role of humans within the increasing robot autonomy levels is particularly important because it is unlikely that given the progress in AI for healthcare (Yu, Beam, & Kohane, 2018) 'most of the role of the medical specialists will shift toward diagnosis and decision-making' (Yang et al., 2017). For now, in this book, we laid down the basis for a rich understanding of what healthcare robots are and how AI empowers them in the hope that the conversation gets started among society.

REFERENCES

Boucher, P., Bentzen, N., Lațici, T., Madiega, T., Schmertzing, L., & Szczepański, M. (2020). *Disruption by Technology. Impacts on Politics, Economics and Society*. European Parliamentary Research Service. Retrieved from https://www.europarl.europa.eu/RegData/etudes/IDAN/2020/652079/EPRS_IDA(2020)652079_EN.pdf.

Carr, N. (2011). *The Shallows: What the Internet Is Doing to Our Brains*. WW Norton & Company, New York City, New York.

Fosch-Villaronga, E. (2019). *Robots, Healthcare, and the Law: Regulating Automation in Personal Care*. Routledge, New York City, New York.

Fosch-Villaronga, E., & Drukarch, H. (2021). *On Healthcare Robots: Concepts, Definitions, and Considerations for Healthcare Robot Governance*. arXiv preprint arXiv:2106.03468.

Gruber, K. (2019). Is the future of medical diagnosis in computer algorithms?. *The Lancet Digital Health*, 1(1), e15–e16.

Riek, L. D. (2017). Healthcare robotics. *Communications of the ACM*, 60(11), 68–78.

Yang, G. Z., Cambias, J., Cleary, K., Daimler, E., Drake, J., Dupont, P. E., … Taylor, R. H. (2017). Medical robotics – regulatory, ethical, and legal considerations for increasing levels of autonomy. *Science Robotics*, 2(4), 8638.

Yu, K. H., Beam, A. L., & Kohane, I. S. (2018). Artificial intelligence in healthcare. *Nature Biomedical Engineering*, 2(10), 719–731.

INDEX

Printed in the United States
by Baker & Taylor Publisher Services

Printed in the United States
by Baker & Taylor Publisher Services